Organised Wellbeing

Wellbeing is now at the top of almost everyone's agenda and many technical books have been published on the topic. More than that, an ever-increasing number of organisations are understanding that getting wellbeing right is a win-win that can boost profits or simply make sustainability viable.

Organised Wellbeing: Proven and Practical Lessons from Safety Excellence seeks to present, in a user-friendly way, all of the key wellbeing themes. It views these through the lessons learnt from safety excellence because, at present, UK safety is world class but, frankly, wellbeing seldom is. In any organisation, culture is even more important than strategy and tactics, and training is just the base of a process in which facilitation and embedding of key behaviours and mindsets is the essential element. Practical and coordinated processes, not initiatives, are required. This book, therefore, seeks to show how aspects of wellbeing, both organisational and personal, are inexorably interconnected.

From an organisational perspective, approaches need to address the truth that 'good work is good for you'. This book, also an individual guide to thriving with passion, compassion, humour and style, is essential reading for health and safety, occupational health and HR professionals at all levels. It is also highly recommended for all managers and staff who seek to maximise their potential and that of their colleagues.

Tim Marsh formed Ryder-Marsh Safety in 1994 and has worked as a consultant in the fields of safety leadership, safety culture and organisational culture generally with more than 400 companies worldwide. He is now the MD of Anker and Marsh and Honorary Professor at the University of Plymouth.

Louise Ward is a Chartered Health and Safety Professional, with over 17 years' experience in a variety of sectors, including nuclear power, newspaper production, investment banking, facilities management, manufacturing and the Civil Service, railway operations and waste water management.

Organised Wellbeing

Proven and Practical Lessons
from Safety Excellence

Second Edition

TIM MARSH AND LOUISE WARD

Routledge
Taylor & Francis Group

LONDON AND NEW YORK

Second edition published 2019
by Routledge
2 Park Square, Milton Park, Abingdon, Oxon, OX14 4RN

and by Routledge
52 Vanderbilt Avenue, New York, NY 10017

Routledge is an imprint of the Taylor & Francis Group, an informa business

Cartoons: Chris Wildt and Roger Beale. Written by Tim Marsh.

First edition published as *A Handbook of Organised Wellbeing* by British Safety Council 2017

British Library Cataloguing-in-Publication Data
A catalogue record for this book is available from the British Library

Library of Congress Cataloging-in-Publication Data
A catalog record has been requested for this book

ISBN: 978-1-138-36842-2 (hbk)
ISBN: 978-1-138-36843-9 (pbk)
ISBN: 978-0-429-42923-1 (ebk)

Typeset in Avenir and Dante
by Florence Production, Ltd, Stoodleigh, Devon, UK

For Kristina, who had only just begun to thrive, and for her friends Georgie and Lauren who we very much hope will do so in her honour.

Contents

Prologue: a week in the life of Joe Egg

Life has been good for Joe recently, and he considers himself really lucky. He has a technical job that has always made the best use of his skills and which, since the buyout by Pine Forest, he now actively enjoys in all respects. What he does is a bit niche these days and not readily transferable, but he'll be in employment for a while at least.

The office he works in is open-plan to encourage sociability, networking, communication and team problem-solving, but with enough screens, large plants and high-winged 'privacy booth' chairs that get the balance right. Walls have relaxing country scenes on them.

Or, 'All that b>*cks', as Joe first described it to his friends.

Pine Forest realised that the company they'd bought had excellent technical skills but terrible morale. A survey they commissioned showed that empowerment, discretionary effort and 'above-the-line behaviours' were close to zero.

They nibbled at this stuff at first. Everyone had personal resilience training a few years back. People liked it and it helped a bit, but Pine Forest have got into the habit of actually following up and checking how effective changes and initiatives are proving. 'A little' wasn't considered a success, so someone conceded that the training-only approach had proved a false economy and then took the plunge and invested in doing it properly.

After 20 years of working for Scraggs & Co. before the buyout, Joe and his colleagues still find this sort of joined-up thinking rather disconcerting!

His new boss is a delight to work for. She sets high standards and challenging goals, it's true, but always in discussion with Joe, and she always checks back to ensure Joe has the support and resources he needs to hit the SMART (specific, measurable agreed, realistic and time-set) targets they agreed upon. Communication with her is always adult-to-adult, and Joe's views and opinions are always respected, even if sometimes they aren't agreed with. He can live with that. Joe's boss has been on excellent assertion, feedback, coaching and motivational interviewing courses, and the learning from these was followed up by in-house coaching – more a mentor, really – and embedded. She now applies these skills by habit and is forever catching him 'doing something right' and asking him, 'What do you need from me to help you work productively and safely?'

Flexitime and the healthy options in the canteen came in straight away, but recently they've even introduced a 'dogs allowed' policy, as it helps spark conversations and is stress-busting – and has really perked the place up. It's saving Joe a fortune in dog walkers. This is in addition to the on-site gym, which provides regular free yoga and meditation classes before and after work.

Joe initially dismissed the meditation classes, quipping that concentrating solely on the movement of air while breathing in and out – or the feel of wood on your bare feet as you very slowly walk about – is difficult when you're also currently mindful of what might happen to your job if a large American competitor buys you out! But he looked at the data circulated after a workshop about how well it worked, tried it and found it worked for him too.

They had a one-off 'financial savvy' workshop there last year that also really helped. Joe knows his technical field inside out but was never any good with numbers and cash flow, and those little plastic cards were getting a bit radioactive in his pocket. Some practical advice really helped. He can't believe how useful he found the 'alertness-building' workshop, or that he went with it to his boss then met up about it afterwards and agreed some changes. An inconceivable turn of events in the old days! He couldn't believe the research findings they came out with or how useful he found the practical tips around sleep, caffeine, hydration and sugar intake.

He knows he's preparing for a good night's sleep from the moment he wakes, and though he still drinks coffee, and too much on a bad day, he alternates with water and decaffeinated, and never 'full-fat' after 2 p.m. He knows to top up his sugar levels with an apple or an orange and not a biscuit or even a glass of orange juice.

Actually, Joe doesn't often use the gym, as he still plays veterans rugby, which, apart from his family, is his lifelong passion. His manager, Helen, has just signed off for him to use his two weeks' volunteer work to help run a rugby camp for disadvantaged children – paid leave to volunteer, another thing that Pine Forest encourages – and he can't remember the last time he was looking forward to anything quite as much.

When the weather's good, he's got into the habit of following his boss's example and taking 'walking around the ornamental lake' meetings with his team. This really helps him hit his 10,000 steps a day that he monitors with the tracker provided to him free by the company. He can't let the department – or the charity that benefits from their initiative – down. The two best decisions he'd taken that year were both made while sitting on a bench with his team, feeding the ducks at the end of a walk. Winnie the dog got some exercise, too. Luckily, his team like him, and that incident with the overconfident duck and Winnie has never been mentioned!

When the weather's bad, he makes use of one of the treadmills in the office – specifically when his brain is in meltdown and no matter how hard he stares at the computer screen nothing happens. This is often just after lunch, regardless of the light and healthy options he's chosen. He knows it's an age thing. At first, like almost everyone else, he thought these treadmills something of a joke and wouldn't touch them with a bargepole, but one day he found he had a brain like sludge, three important calls to make and was a few thousand steps short of his target, so he gave it a go. Thirty minutes later, he'd made three excellent calls, had caught up with his daily steps deficit and felt really quite refreshed, so he decided that maybe they weren't such a stupid idea after all. The trouble is everyone had started using them and they weren't always free any more.

When that's the case, he recalls that YouTube clip he watched after the alertness workshop by that ludicrously healthy-looking 60-year-old £5,000-a-day 'energy guru', finds a quiet spot and knocks off 30 sharp press-ups and 30 star jumps.

Last week at the annual conference the theme was 'learning' and they had a guest speaker. This chap was a professional horse rider who had lost more races than any horse rider in history. More than 14,000. And he'd fallen off hundreds of times too – broken almost every bone in his body. The twist was that he wasn't an Eddie the Eagle type but Tony McCoy and he'd also won more horse races than anyone in history, more than 4,000. His talk perfectly complemented the theme of the day. This was that it is impossible to be error free, especially when striving for excellence, so strive to avoid arbitrary blame as it's death to a strong culture and excellence generally. At his next coaching session Joe found himself shouting not 'ughhh' and 'awww' when the lads dropped the ball trying out a new move but 'nearly, *nearly*. . . really nice try . . . I'm liking that ambition'.

After work, Joe meets his mate Mike down the pub for a couple of quick ones. He's started to make a point of doing this once a week, ever since he went on a hygge workshop. This is the 'new big thing', apparently – a Danish concept (books on it in every airport bookshop) that apparently underpins them being the happiest people in the world. The basic principle is that a little of what you fancy does indeed 'do you good' and that a couple of drinks with friends in a relaxing environment can be a key part of a balanced and all-round healthy lifestyle. The woman who ran the workshop explained that blokes especially need to physically do things with other blokes to maintain friendship and that, like everything else in life, the grift principle applies, and you get out what you put in.

She also explained that meaning and giving in life are key for most people. Joe pondered on this for a while and subsequently started coaching under-12s rugby on Sunday mornings. She was right; it's the highlight of his week. In the old days, he was too hungover to even turn up and watch his son play on a Sunday, but those days are few and far between now. Indeed, he often has a nice little 'Sunday morning chat' with his wife before setting off. It makes a change from the old days when she'd shout at him for being a 'useless drunken git' when she had to cover for him taking his son to play – but then that was back in the dark old days when he still worked for Scraggs & Co., and he was, quite frankly, 'really totally and utterly . . . *very stressed*'.

The first time they went out was after Joe picked up the phone and told Mike he'd been to a workshop where they explained that men in particular need to do things together to maintain friendships. And that since Mike had stopped playing for the vets rugby team, they'd not seen much of each other. After a few games of pool in their old noisy and aggressive pub, they move to the new bar they used to dismiss as 'too posh' and take a comfy seat in the subdued lighting and talk for a bit.

Mike needs the chat, as he's having the same family troubles that Joe had a few years back, but complains, 'We don't get employee assistance at all – let alone that CBT-based family therapy you got!' Joe recalls that two sessions in, he still thought it a 'load of old b>*cks', but that after the full course of six sessions things had actually improved immeasurably. On their second glass, they, as they often do, talk briefly (if reasonably discreetly) about their sex lives. Joe commiserates with Mike, and Mike says that the relative improvement in Joe's relationship with his wife in this respect in recent years is quite appalling to him – commenting that it's, quite frankly, *obscene* that a couple who've been together as long as they have enjoy a decent sex life.

Mike is a safety expert, and, moving the conversation on when it's apparent Joe's happy to continue to boast about his sex life indefinitely, points out that he thinks Joe's company recently saved their lives. Mike explains that the defensive driver course Joe went on was really useful, of course, for all the pro-active driving techniques it covered, but that he's just read a research paper that explained how a person's peripheral vision works several times better when a person is relaxed. He explains: apparently 'fight or flight' is great for an obvious threat, in the short term, but you really want to be in 'rest and digest' when a truck comes out of nowhere like the other day. He concludes, 'Honestly mate, I really think that if I'd been driving instead of you, I'd not have seen it and we'd have been toast.' Mike never drives these days, as he's got into the habit of starting early with the beers and isn't so good at putting the cork back in the bottle.

With that, Joe's wife comes in looking healthy and happy, and, to Mike's incredulity, pleased to see Joe.

They part at the door, Joe to go back home now he's been 'tagged', but Mike announces that maybe he'll pop back to their old boozer to see if anyone fancies a frame or two. 'Just for a quick one.' As he sets off home, Joe offers, 'Have one for me, but . . . hey . . . look after yourself Mike.' He notices that Mike stops to light up a cigarette and that, on top of everything else, the daft f**k has started smoking again.

The next morning, Joe sees Mike coming out of the greasy spoon café and heading for the bus stop, and stops to offer him a lift. The quick one turned into quite a few slow ones, admits Mike, and because of this, his wife really wasn't best pleased, so he spent the night on the sofa. He admits, 'I'm nowhere near under the limit yet, and just can't lose my licence on top of everything else.'

They discuss that it's not great that the massive fry-up that Mike just polished off wasn't stomach-turning but is now what a hardened drinker would call 'good soak' to Mike. They also agree that it's a good job that Mike now works in an office rather than on the tools still.

The next morning, Joe sees Mike outside the café again, but this time he's screaming into his mobile phone. Some people passing are finding the stream of foul language and creative threats of violence amusing, even when the phone gets thrown at the ground and shatters into a hundred pieces. Many others are looking alarmed. 'Bad news?' asks a worried Joe, but Mike, looking sheepish, admits, 'No, just a f>ing cold-caller.'

They both laugh. Joe calls Helen to check it's OK if he's an hour late and then takes Mike for a coffee. Their relationship has been based on blokey banter and taking the Michael out of each other, and Joe has just been on his mental health first aider course and knows what to do. He knows that talking honestly to a peer that you trust is even more effective than talking to a trained counsellor and that it's clear that he's the peer in question. As they get the coffee in there's still lots of banter about the phone incident. 'Good job that bloke was only on the end of the phone!' But when he gets serious, changes tone and asks Mike how he's really feeling, Mike wells up, then his shoulders start to shake.

They talk a while, then Joe gets his phone out and forwards Mike a whole list of contact numbers he might find helpful. They look at each other. They laugh. 'Maybe I'll write them down,' says Mike.

The next day, Joe is called into Helen's office. A leader like her is in great demand, and she's approached by headhunters all the time but has always said, 'Thanks, but no thanks.' Until today, that is, as she explains to Joe that Pine Forest are doing so well financially they are being bought out by Mega Bucks & Co., who have a worse reputation than Scraggs even. Joe blows out his cheeks and Helen shrugs back with a sad smile and as much empathy as she can muster. They both know what's coming.

She has a call from a headhunter to return. Joe rings Mike and asks, 'You know we talked about you cutting down on the booze? You don't fancy delaying that by a day or two, do you?'

Introduction

1

A handbook of wellbeing by one of the world's leading safety organisations and someone known for their work in safety culture and safety leadership? At a glance, you might ask, 'What that's all about?'

Some context:

- Britain is a world leader in safety. Its figures are outstanding and its expertise is sought worldwide.
- Britain lags behind in wellbeing, even though James Tye (who founded the British Safety Council and who inspired the HSW Act that underpins much of the UK's basic excellence) was an early advocate of wellbeing.
- Safety is, of course, a part of wellbeing – indeed, a key element at the sharp end!
- We think it's really clear from where we sit that many of the hard-won lessons of safety excellence are not being replicated in the wellbeing field (something the Campbell Institute White Paper also says of the US). Here, as in the US, wellbeing initiatives often look to be making many of the same mistakes we were making in safety 25 years ago!

These lessons being, among others:

- Education and training are but base one, because on a day-to-day basis 'temptation' (or ABC theory) kicks in, and we are apt to act in a short-term way if the environment isn't supportive.

- We've learnt to target the 'back of the room', as unless we make a concerted effort to effectively target those that most need it, we just subsidise the 'already well' with apples, gym membership and step-counting watches they would have bought anyway.
- Objective analysis and targeted facilitation are key to behaviour and culture change.
- Initially, an organisational, not an individual, focus is key.
- Nothing is more important than genuine senior management buy-in.
- Lessons from the world of behavioural economics (or nudge) research should be utilised by SHE.
- The win-win case must be made convincingly to engage all stakeholders.
- An interlinked and holistic approach that is both bottom-up and top-down is essential.

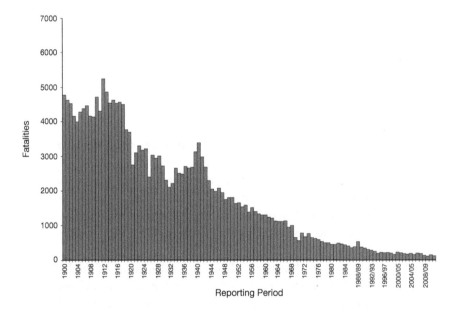

It's clear that substantial improvements have continued decade after decade. If we ease off we'll drift backwards, but massive improvement from here simply isn't possible. Three other points are worth making. The first is that figures for working people who took their own lives last year would fit in perfectly with the 1900–1920 figures to the left. The second is that in recent years almost twice as many schoolchildren took their own lives as people who were killed in work accidents. The third is that the number of people who died as a result of exposure at work last year (to quote the deliberately conservative estimate from the Rushton Report) is over three times the 1900 to 1920 fatal accident rate illustrated here. Debate continues as to whether things are actually getting worse.

> In terms of wellbeing and productivity . . . Both the UK and Iceland are around 30th in the world . . .

At 0.8 the UK is bested only by the small country of Iceland. Comparing it to other countries of a similar size and demographic we see that France, Germany and Japan are around 3, the USA just under 5. (Note that many developing countries are at 20 or more).

This book therefore sets out to write a practical guide to the win-win that is wellbeing excellence. We're world leaders in safety so we really ought to be world leaders in wellbeing, because there's far, far more at stake both ethically and financially. Much of this book will be about the likes of Joe in the prologue, and about what we all need to thrive rather than merely survive. First, however, it's worth considering the massive cost to organisations and families regarding those who are struggling to get by.

The UK young persons' mental health service (CAHMS) is officially in crisis according to a recent White Paper, and the number of young people struggling with mental health (MH) issues appears to be increasing at an alarming rate.

The problem is prevalent among adults, too, with figures published by the mental health charity MIND indicating that one in four people in the UK are affected by mental illness in any given year. ONS statistics show that suicide is the biggest killer of men under the age of 45, and studies in the construction industry indicate that male workers are six times more likely to die through suicide than through a fall from height.

Stories about these issues are all over the media; this suggests increasing numbers of staff are struggling with their own mental health or are distracted by family issues, which means organisations need to consider this as part of their risk profile. When, using average rather than worst-case figures, a fifth of people are struggling and the trend is towards a quarter, then UK plc is clearly going to suffer.

We might hope that the government has a coordinated strategy to deal with these issues, but a glance at the papers will tell you how well that's going. There may be the will and there are warm words, but there's no money. We look to have a neo-liberal (and Brexit-distracted) government in place for some time. Any significant improvements in the next decade or so are going to have to come from business. The good news is that business has a track record of being quicker to identify trends that pose a risk to survival and of coming up with proactive solutions to mitigate. In fact, for right or wrong, that is the mantra of the currently dominant neo-liberal socio-economic view ('Get government out of the way and let us deal with it').

Proactively identifying and tackling risks to sustainability is what good businesses do, and wellbeing is at the core of a sustainable business.

Most people spend as much time, if not more time, at work as they do with their families. Their interaction with their employees and colleagues is utterly central to any coordinated approach.

Successful wellbeing programmes are often referred to as the ultimate win-win: good for the individual and good for the company. We'd like to suggest a third win: good for society as a whole.

With that third win in mind, the Erasmus project in Spain, Iceland and Poland is encouraging the over-50s to 'train' for their retirement. In 20 years or so, the over-65s will make up a full 25% of the UK population. Hopefully, we'll be fit, healthy and will have experience that could be channelled into studying those subjects we always promised ourselves we would, and volunteering and/or coaching and teaching.

There are two reasons why organisations might want to support such schemes. The first, of course, is they may develop a cheap resource, and the second is that presenteeism is an ongoing issue. This can be especially true of the older worker, with little chance of further promotion, who is coasting to retirement. These people do not always lead by example as we want them to, but, on the other hand, direct action against them will seem heartless and unfair and will have negative consequences for morale.

In this age when senior managers come and go quickly, these experienced, time-served mid-management staff are key. They are the organisational 'memory' and they have a huge role in determining the day-to-day culture and tone.

Surely, energising them cannot fail to make business sense?

Again, the interconnectedness of the wellbeing field is illustrated. We need to feel that we are contributing. Empowerment programmes deliver this directly, and Erasmus-like programmes engender a mindset that 'I'll be contributing for a while yet'.

The Campbell Institute White Paper sets out the challenge well. It summarises that the best organisations they talked to (from a self-selecting sample of the best of the US organisations) have good programmes that 'have yet to become excellent wellbeing systems'.

For us, an excellent wellbeing system is one that is holistic, coordinated, user-friendly and systemically embedded within a culture of care and health.

Background and context

Traditionally, focus has fallen on safety rather than health issues in the workplace. Huge progress has been made over the last 40 years, and the number of fatalities and serious injuries has reduced significantly. The economy has changed too, placing different demands on workers, and as the number of accidents has decreased, there has been an increasing awareness of longer latency health issues.

Today, skilled and knowledgeable people are at the heart of the modern British economy. Employers recognise that their workers are a key asset and are keen to support and retain them. Brexit has sharpened this focus too, particularly in sectors such as healthcare, hospitality and construction, where there is currently a dependence on skilled labour from other EU countries. There has been a blurring of the boundaries between life and work, as technological advances – and the changing nature of work – mean that many people can now operate effectively from any location with an Internet connection. This can result in excessive working hours, but also allows people to work much more flexibly to accommodate personal and family requirements. The gig economy is changing employment relationships too, and an increasing number of people work remotely, outside of a traditional workplace, which can impact negatively on mental wellbeing.

The time is right to focus on health as an enabler to work. However, it's not just about employing and retaining people with disabilities or long-term health conditions. There is also a need to focus on health and wellbeing for the existing working population. People are living longer, and pensions are frequently no longer sufficient to support early retirement. As a result, the workforce is ageing, and many more people are having to manage the health effects of getting older while still at work. People need information and support to help them make healthy decisions, and timely access to medical services is a key requirement. Short-term illness and minor injury can easily result in long-term sickness absence if the issues are not addressed promptly. This has negative effects on the physical and mental health of the individual, results in costs for the employer, and places demands on the medical and social care systems.

There is a need for more openness and effective engagement between individuals, employers and medical and social care services, supported by revised and streamlined systems that recognise work as beneficial to physical and mental wellbeing and that enable effective employment.

There is no doubt that the right work delivers positive benefits, both physically and psychologically, for individuals, but also for the employer and for society in general too.

What's holding us back?

Scientific and technological developments mean that we now know more than ever about how our bodies and our minds work; how they can go wrong, and what we can do both to keep ourselves healthy and to treat illnesses and injuries when they occur.

For the last 40 years, we have focused on keeping people safe at work, and we've seen remarkable improvements, but the management of health at work lags way behind. The business case is clear, so why aren't we better at it? Could it be our very 'Britishness' that is holding us back?

Research suggests that British people are far less confident about their bodies than their counterparts in other areas of the world. On holiday, it tends to be the Brits who are covered up by the pool or changing under a towel, and at work we are equally inhibited. But this attitude is preventing us from managing our biggest asset effectively.

We all need to get more comfortable talking about health. This isn't a static thing. There are lots of different factors that affect our physical and our mental health in both positive and negative ways. We all need to get better at recognising this and reaching out for support when we need it.

We need to work on establishing a positive wellbeing culture in our workplaces. Work can be really beneficial for both physical and mental health, but there are also times when we need to take a break or get specialist support in order to let our bodies and our minds recover. Every case is individual, and the best way forward is for employers and employees to feel comfortable discussing issues as they arise, and building a plan together that ensures that the right action is taken at the right time.

Peer support is really important too. People have great capacity to help and support each other, but we need to move away from the traditional British stereotype and the 'stiff upper lip', 'grin and bear it' type of culture if we are to support each other effectively. It's normal to feel unwell sometimes, and when we do it's good to ask for help! Recognising this will help to develop an engaged working culture that will build resilience.

How can we move forwards?

There is a wide range of factors that can influence physical and mental wellbeing, and as the issue is gaining recognition, we are seeing an increasing range of products and services being marketed to employers. The options may seem endless, but resources are limited, so how can employers ensure that they use the available resource to the best effect?

Currently, many organisations go for the easy options – fruit at the tea point, healthy options in the canteen, discounts on gym membership and/or a cycle-to-work scheme – but without any needs analysis or measurement, they have no real information about whether this is actually improving wellbeing.

A more targeted and analytical approach is required to ensure that a proper benefit is realised.

There is no 'one-size-fits-all' approach to wellbeing. To be effective, initiatives have to address issues that are of relevance in a particular workplace, or to a particular group of workers. Data held within the organisation can help with this, but it's also important to engage with workers right through the organisation. They have a wealth of knowledge about issues and problems, but they will also have ideas about solutions too, and they will be far more likely to engage with wellbeing initiatives if they are involved in developing them.

The business case   2

In these days of social media, with everything archived and accessible, a lack of authenticity is very difficult to hide, and through social media and employment forums there has been a definite blending of companies' 'retail brand' and their 'employment brand'. A business needs to address its corporate social responsibility (CSR) obligations well enough to avoid bad publicity while simultaneously avoiding alienating and damaging their workforce to the extent that they go online about it in numbers. It's increasingly difficult to survive, let alone thrive, as a business with a bad reputation, especially when vicious rather than virtuous circles kick in. Schneider's famous ASA theory says that certain organisations will attract certain people, will select in their own image, and will in time reject or be rejected by people who don't fit.

If you're lucky enough to get a good applicant, you may well select them, and of course won't reject them when they start, but they're very likely to reject you, and Generation Y are becoming infamous for their 'Give me three good reasons why I should even think of staying' mindset.

Rather than just damage limitation, however, there seems to be increasing agreement that a 'happy' employee is around 10–12% more productive than the average worker, with an 'unhappy' worker around 10% less effective – giving a spread of around 22%. Soma Analytics found in 2017 that FTSE100 companies that prioritise engagement and wellbeing outperform the rest of the FTSE by 10%. These percentages reflect huge monetary figures. The Stoddart Review, for example, suggests that for every 1% improvement in productivity, UK plc sees an increase in around £20 billion in national output. A 2016 CIPD study suggests that stress and mental health costs UK business around £35 billion a year.

In a global ranking of wellbeing, health and happiness, the UK comes in at 18 out of the G20. The 'Glassdoor' survey found that a one-unit increase in its 10-point rating scale equated to a 7.9% increase in market value. They also found that the increasing awareness of this issue has led to talk of a '2nd P&L account' to sit alongside the one most often looked at by shareholders. The second account covers 'People and Leadership', and very few companies can quote figures for that. They should be able to, though, as it's not rocket science to go out and assess just how disillusioned, or not, your workforce are. Any sort of systemic and replicable method of assessment will do.

Absenteeism and 'presenteeism'

Studies suggest that for every day lost to absenteeism, around 9–10 are lost to presenteeism. Figures start at around 0.6% lost to absenteeism (versus 7.4% to presenteeism) in the best organisations, but more typically companies will lose between 0.7% and 3.7% to absence and between 18% and 31% to presenteeism.

Turning up for work when ill enough to stay home may feel like a sign of commitment, but it correlates with negative issues. Other than simply infecting co-workers and the financial imperative behind zero-hours contracts, it may be that the organisation is so lean that all hands are needed at all times, and everyone knows it – or that communication with management is difficult and infrequent. There may be concern about 'being out of the loop' or a perception that blame will follow an absence.

In short, all the experts agree that employees turning up for work but who are unproductive is a clear warning that something is wrong with that culture.

The interlink with the historical P&L is, in the medium to short term, far more important than the 12% 'effort' deficit on any given day. It's far more nuanced and multifaceted than that, and a bit like the scene in Monty Python's *Life of Brian*: 'What have the Romans ever done for us?' Engaged and motivated workers:

- work a little harder; and
- leave less frequently (and it's always the best and most attractive workers who go first).

But they also:

- (need to) take less time off sick;
- (genuinely) recover from illness and injury more quickly;

- prove better trainees; and
- contribute better to team problem-solving.

They are pretty much better at everything. They are even far less likely to die after a first heart attack or have a second heart attack.

It's been estimated that for every 10 workers off with a 'bad back', eight are off because they're somewhere between 'fed up' and clinically depressed. Linked to this, positive people succumb less often to mental health problems. Demonstrating the sort of savings involved, Severn Trent demonstrated that a 1% reduction in sickness absence from 4.3 to 3.3 meant a £1 million a year saving.

In the US especially, generous health insurance is a key attraction for employees, and the premiums the company pays, based largely of course on payouts, are a significant percentage figure. Savings here go straight into the profit margin, and more and more companies are acutely aware of this opportunity. The Campbell Institute define an integrated system as covering both safety and wellness, with this including injury and illness as well as wellbeing, covering physical, mental, emotional, social and economic health: 'The latest focus for maintaining a sustainable business enterprise has moved from beyond workplace safety to include employee health and wellness.'

The Harvard Business Review describes successful wellbeing (WB) programmes as being aligned with overall company identity and goals, having engaged leadership, being comprehensive in scope and quality, and easily accessible in terms of access and cost. As well as agreeing with the above lists (indeed, because of the above lists), the Harvard Business Review study suggests a return on investment of '300% or more'.

Companies on the Harvard Business Review 'Best Places to Work' list outperformed the market average by 115%.

A key lesson from safety

We've often talked about halving a problem (usually lost time accidents, sometimes fatalities, and sometimes mere near misses) as a 'step change'. Many, many companies in the UK and around the world have achieved this in a year to 18 months and then repeated that success. Indeed, several industry-wide initiatives have been labelled as successful 'step changes'. It's not unrealistic pie in the sky – it's achievable.

This is good, as simply halving the gap in engagement between the UK and the G7 average would be worth around £200 billion!

And, to reiterate how achievable that is, we're not talking about achieving excellence and being the best here – that's merely getting the UK to a place that's only half as bad as it is now! In short, even if this issue is addressed entirely for ethical reasons, that's one hell of a war chest to pay for it.

Case study: building industry

A UK builder described at a conference how their industry was bedevilled by a high churn of staff because an offer of £2 an hour more was enough to induce a worker to move on. After the introduction of a systemic empowerment and enrichment approach, turnover decreased significantly, because now £2 wasn't enough to induce an electrician or a bricklayer to walk away from a job they now enjoyed. (The key here, it was suggested, was the installation at workers' camps of soundproof rooms with excellent Skype connectivity that could be booked in 20-minute blocks by migrant workers for conversations with family at home.)

Case study: from the frontline

The UK care industry is in utter crisis. Private investors are leaving the market in droves, and public providers are withdrawing services across the board. Around 900 staff are leaving the sector every day. Bucking this trend by taking a staff wellbeing-centred approach to sustainability, Sandwell Community Caring Trust have two key indicators: staff turnover and sickness.

As well as training, staff are paid as much as is viable, given extra holidays and benefits, and are consulted and involved in the governance of the business. The reduction of turnover from 40% to 10% means recruiting 200 fewer staff a year, and reducing sickness decreases the reliance on (more expensive) agency staff. These time and cost savings are focused on wages, training and asking the cared for what they think of their service.

The CEO, Geoff Walker, says that it makes a low-margin business sustainable – the impossible just about viable. The focus has to be on the quality of care, and where 15-minute social care visits from badly motivated and ever-changing staff is increasingly the norm, it enables perhaps the most important key performance indicator (KPI) of them all: an increase in quality interactions and friendship.

These interactions are between some of the most vulnerable people in our society and some of the mostly badly paid and over-worked employees.

Problems organisations face implementing WB programmes

The question is begged: Why isn't everyone doing it, and doing it well? Part of the reason, of course, is that it isn't as immediate as safety, so it's difficult to see the damage that's being done and the opportunities lost.

With home life and work life ever more blurred, the demarcation is even less clear than it was. The question is begged: What's causing what, and who's responsible legally and/or morally? The days of patting someone on the back at 5 p.m. and feeling satisfied they're going home with the limbs and eyes they arrived with are long gone. Few people can now leave work entirely in the factory, and even fewer their cares. Advances in organisational complexity and enhanced communication devices were always likely to correlate negatively with stress.

The experience of occupational health shows how this works. In the UK last year, a conservative estimate (Rushton Report) is that some 13,000 people died of occupational exposure, and typically from exposure decades ago. This is often exacerbated by smoking and drinking but also by air pollution and stress. Though asbestos and other issues are now well known and controlled, many new risks are entering the market with nanotechnology, and the experts suggest that the 13,000 figure is likely to increase. The problem is that the politicians and CEOs in charge today will have long since retired when the price needs to be paid. We have come a long, long way in addressing the S in SHE, but there is still a long way to go towards the H.

Specific problems regarding the effective cooperation in the design and delivery of a coordinated and holistic approach to the wellbeing of staff, articulated by the Campbell Institute and others, include:

- It (typically) needs HR and SHE to cooperate, or at least for HR to have the practical field experience a good-quality SHE department will have.
- It can run headlong into union suspicion (and/or a turf war).
- The people it most needs to target (middle-aged men) are the people hardest to reach.
- Tracking progress can fall foul of data privacy issues.
- Initiatives fail to reach the people that most need them.

Coordination between SHE and HR is vital. One example of how not to coordinate is the major employer who rolled out some behavioural standards through HR that included reference to six core values, including 'safety', without

What Has This Wellbeing Ever Done For Us?

"OK, fewer of our best staff leaving. A reduction in absenteeism and presenteeism. More creativity, discretionary effort, productivity and a better reputation ... granted ... but apart from that ..."

consulting the SHE team in any way. They therefore struggled to support their colleagues, not just because of legitimate 'buy-in' issues but because the definition of said safety behaviours was, to quote the head of safety, 'crap'. In any decent culture, the SHE team will have an undiluted 'here to help' credibility that a multifunctional HR team might struggle to match because the HR team will have to lead discipline and the communication of bad news. (In a really poor culture, the SHE team may be seen as an 'elf and safety' inconvenience, of course).

Turf wars between HR and SHE are possible at the best of times, and the same is true of unions. One of the authors had a senior union official once say, 'If there's any empowerment work to be done with my members, I'll decide when and how.' Things are rather different these days, but the writer Gareth Morgan sums it up well when he suggests that resources and decision making is power, and resources are always scarce, so people and departments will be inevitably be disposed to compete for them, and politics is inevitable.

The solution, as ever, to turf wars and other politics, is strong executive leadership, and this is the key reason why executive buy-in to a holistic plan is so important. It's just a variation on one way Daniel Kahneman describes what he means by thinking fast and slow. He suggests if a major change will work well for 7 out of 10 companies in a group but fail for 3, then no one individual company will ever choose such odds. This needs to be an executive decision, properly coordinated and supported.

The same is true with wellbeing.

Mental health

We all have mental health just as we do physical health, and it is equally variable. There is a huge range of factors that can impact positively, or negatively, on our mental wellbeing, and it can have just as profound an effect on our bodies as physical illness does. The difference is that there is a stigma attached to mental health; it is a social taboo that makes people ashamed to talk about it or seek help. This is compounded by the lack of obvious physical symptoms. If someone has a streaming cold or is limping when they walk, others will notice and (hopefully) offer sympathy or assistance. The trouble is that you can't see when someone's mental health is poor, so a great many people are left to suffer in silence, and this can have tragic consequences if it isn't addressed.

Suicide

More than 6,300 people killed themselves in the UK during 2015, and around 70% of them were in work. The figures skew greatly towards middle-aged men and people in poorly paid work. Risk factors include sensationalist media coverage, childhood trauma, problems with mental health, issues regarding the misuse of drugs and alcohol – where 'self-medication' may be involved – and, of course, relationship problems.

So a typical suicide, if that's not an insulting generalisation, might involve a person who gets little satisfaction from work or who has never really had an emotionally literate role model when young or learnt to talk about feelings and issues. If the one person, typically a partner, who they may have opened up to pushes them away, then they may feel that there's no one left. A friend might later comment, 'We had a typical Friday night out. We had six pints, a game of snooker and a good chat about the football. He seemed his normal self.'

What organisations such as the Samaritans stress is that often a simple conversation about how they feel is all that's needed to give pause for thought. The Samaritans alone have more than 5 million contacts a year, but a colleague is well placed to proactively notice there may be a problem and initiate what might be a life-saving conversation. They also stress that mentioning suicide specifically is not the 'worst thing we can do' as many instinctively feel but may open the door to a vital conversation. They stress listening is all that's required and that what we shouldn't do is reassure them 'it'll be ok', as we can't know that. Nor should we talk about personal experience ('I know just how you feel' . . . 'You remind me of Uncle Arthur'). We just need to ask open-ended questions about their feelings and listen.

More specific initiatives with links to the Erasmus project mentioned in Chapter 1 include that of 'men's sheds'. Here, men wishing to learn skills can be taught them by others who, frankly, may well have little better to do with their time and who welcome the chance to pass on the skills they've acquired.

'I'm fine, thank you': the problem with men

A basic 'walk and talk' tour of the shop floor will, 9 times out of 10, generate the answer 'fine' if the right questions aren't asked in the right way. This is especially true of men, who can be hugely reluctant to open up, even when in the depths of despair. The basic approach is to ask open, high-value questions, not closed, low-value ones. ('Are you OK?' is an example of a closed, low-value question as it has a yes/no answer. An example of an open, high-value question might involve asking someone what they think or how they feel.)

Dealing with people (especially middle-aged men) may look, at a glance, to be complex and intractable but also needs to be tackled proactively.

Several interviewees suggested that facts and figures articulate the win-win only so far, and that stories are needed to personalise and illustrate the point. We need to appeal to emotions as well as the balance sheet. Many a culture change has followed a workplace fatality or suicide that has brought home the reality to a director. This is too reactive, however, and ideally we'll be proactive in our approach to this important issue.

A lesson from safety

Approaching the target audience in the right way is key because knowledge is only base one, and the people we most want to reach are also the ones most likely to dismiss out of hand anything not pitched in the right way. The three experienced blokes that sit at the back of a briefing, arms folded, when being told about the risk of doing X and Y and who then go out and do X and Y are the same as those who sit at the back taking in wellbeing advice with a shrug.

We quickly learnt that if you want to reach people, you have to engage in dialogue in a language they understand and never patronise to maximise the chance of discovered learning. Talking to people like adults and using data and illustration helps maximise the chance of 'discovered learning', leading to a genuinely internalised promise to ourselves – which is the only sort we ever really keep. We'll detail tactics and techniques later, but it has to be addressed at the strategy level. Utilising 'nudge' theory must be part of that strategy. Nudge, or 'behavioural economics', is the science of facilitating individual change through making an adjustment to the environment that encourages more positive behaviour whilst requiring minimal effort on the part of the individual.

The safety world has long used accident victims to drive home the personal cost of an incident, and increasingly the wellbeing world is using the same approach. Ideally, this won't just be people recovering from a breakdown but more proactive testimony. Recently, Prince Harry described how long it took him to come to terms with his mother's death, despite him looking to have coped well.

A global director of SHE, who was once the company doctor, told me that discussing general health issues was easier with people who seemed vulnerable in his surgery and somewhat scared about what he was going to tell them.

Specifically, he mentioned this proved useful when a smoking ban was about to be vetoed as 'too difficult'. Making eye contact with colleagues he'd recently seen in his surgery, he drew attention to the fact that, given the size of the company, it would prevent dozens of people a year from being sat down in a doctor's office and being given terrible news. He's too professional to have asked anyone 'you remember how that feels?' but the point was made and the ban stood.

James Reason coined the term 'safety is a guerrilla war'. In safety, we long ago leant to use what works, and much of the wellbeing work is still very polite, formal and restrained. We need to rediscover the 'at all costs' attitude of James Tye and get stuck in. A practical example: one interviewee said they'd had a fantastic response to a prostate and testicular cancer awareness campaign. They used a professional comedian, who effectively did it through the medium of the 'knob joke'.

Not exactly politically correct. But it worked for the target audience.

A case study in how not to do it

One organisation we interviewed stressed that they had produced an excellent wellbeing webinar that talked employees through key resilience techniques and lifestyle. They'd made a point of keeping it available indefinitely after first broadcast, they boasted, so we watched it. It was all good material, but the webinar was merely optional. Quite aside from the issues that 'education is but base one' considered elsewhere in this book, this meant, of course, that it had been watched a mere 60 times to date.

They employed more than 10,000 staff, and we're pretty certain that few of the 60 were from the population they most needed to reach.

Wellbeing and the individual: a holistic framework

3

We'll of course consider organisational issues in due course, but we'd like to start with a basic holistic model of overall wellbeing as it relates to an individual.

A holistic five-factor model of happiness and wellbeing

1. Good work

Have a job that you enjoy, because good work is good for you (bad work isn't). Multimillion-selling writers such as Marcus Buckingham emphasise that we are unlikely to excel at anything unless we enjoy doing it, as otherwise we simply won't put in the energy, effort and passion to maximise our potential. Of course, economic realities can crash against Hollywood 101 'follow your dream' films here, and there are exceptions, though sometimes they illustrate the principle. For example, reading John McEnroe's biography, *Serious*, it's striking that he didn't actually enjoy his career. He was driven mostly by a fear of failure. He certainly excelled, but this internal conflict definitely showed sometimes! He now has a job that he genuinely enjoys, commentating, and seems a different person entirely.

Organisations can of course help with this element by maximising involvement and empowerment while being mindful not to over-empower to a level that is stressful (we'll walk through Warr's vitamin model of work in detail later). Being involved in a successful process or project, HS&W-related or otherwise, can be hugely rewarding and energising. For example, one of the authors had a safety team volunteer once say, 'Twenty years I've been with this company, and last night [preparing a presentation to be given at a hotel] was the first time I've ever done anything for them on my own time. My wife nearly had a heart attack!'

> 8 Million people in the UK (1 in 3 employees) are one pay check away from not being able to pay their rent
>
> . . .
>
> (Shelter)

He was very nervous about presenting to his board, but he did a good job; later he was propped up against the bar, clearly enjoying a bit of a 'boozy bonding session' with the CEO, who had warmed to this chap's enthusiasm.

2. Financial wellbeing

A recent study shows that, once above the amount required to live on the threshold, doubling your income increases a sense of wellbeing by a score of just 0.2 (on a 10-point scale).

In a recent Kronos study, remuneration did make the top 10 reasons for employees leaving their job. It came 9th.

These stats are very illustrative that money isn't happiness for most people, but it is important that they have enough money. The relationship with money isn't even entirely linear as, for example, a surprisingly large percentage of people (30%) who win the lottery say it made them less happy in time as they handled the change badly. Of course, the better we do with the other four elements discussed here, the less we need! For example, if we give workers a 20% pay rise a propos of nothing, job satisfaction of course goes through the roof . . . for one or two months . . . then it goes back to pretty much where it was before!

Books such as *The Spirit Level* argue that absolute wealth isn't as important to happiness as relative wealth compared with others – which is why observations such as 'We're all in it together' work (but not when delivered by multi-millionaires)!

Even people who think they have this in balance can be very wrong. In short, the following quotation often applies:

> *Many of us are stressing ourselves doing things we don't enjoy, earning money we don't really need, to spend on things we don't really want, to impress people we don't even like.*

Someone recently told me that one in three working families in Britain are just one pay cheque away from losing their home! The statistic comes from research by Shelter, which also reveals that one in three low earners regularly borrow to cover their rent, and that 150 families in Britain become homeless every day.

These sobering statistics illustrate that a significant percentage of the working population are just getting by from day to day. As well as impacting directly on mental wellbeing, such immediate and significant personal concerns are very likely to impact on attention and concentration and could therefore also contribute to accidents in the workplace.

Interestingly, recognising that money worries can impact severely on stress levels and work performance, several organisations have started to offer a basic course on 'financial education'. The way it's made available needs careful handling, of course, but almost all of us could do with being more financially savvy! And for those who are 'useless with money' and struggling, it has to be better than learning lessons from lenders with exorbitant interest rates or the local loan shark!

3. Family and friends

This is a huge and personal area of wellbeing that interacts with all others in a dynamic way. We can't possibly do it justice, but this is a holistic book, and there are some comments worth making for orientation. These are:

- All non-technical skills for maximising the empowerment and potential of your colleagues of course apply equally to family and friends. (Indeed, the guru of positive psychology, Martin Seligman, says that we should have a ratio of 3:1 praise to constructive criticism in work, but 5:1 at home. This translates as: you should always be your child's biggest supporter but not to the extent that you ruin them).
- As ever, the 'grift' principle covered in Chapter 4 applies. Here, as every-where, you tend, on balance, to get the luck you deserve based on the effort you put in. (There are no guarantees either way, of course, and we know many readers will have said something sceptical, and quite possibly extremely rude, out loud.)
- Every day, people are walking in the countryside, talking and laughing with friends or family, and 'splashing out' by buying themselves an 'earned treat' of a choc ice or a coffee half way around the walk. We all know how good these things taste in circumstances like that. Huge numbers of readers will have promised themselves 'more of that next year', especially if they've just

watched *It's a Wonderful Life* again, but then fail to keep that promise. The section on ABC or temptation analysis, also in Chapter 4, helps explain the perennial problem.

Finally, it's worth stating the truth that for the vast majority of people you cannot maximise your wellbeing alone. Try this simple exercise, which seeks to illustrate why we need to keep making the effort, no matter how difficult it is and how annoying other people are. Quickly write down your three or four favourite memories ever. The stuff that we hope we're thinking of on our deathbeds.

For many readers, every single one will contain the word 'with'.

4. Physical and emotional wellbeing

In the world of safety, disabled presenters such as Jason Anker, who was paralysed in a fall at work, and Ken Woodward, who was blinded at work, articulate the importance of maintaining physical wellbeing in the most profound way. The first thing we need to do at work is to avoid physical trauma. Many lives have been blighted by merely seeing a physical trauma, as the Outtakes film *The Witness*, about a colleague of Ken Woodward, makes clear.

Emotional wellbeing is, of course, the main topic of the book, but it's worth reiterating here that any organisation must have safety excellence as a cornerstone of its approach.

There is a huge range of factors that impact both positively and negatively on our physical and emotional wellbeing. It's important to recognise that these things are fluid and will change continuously. This is normal, and the important thing is that we learn to recognise how we are feeling and adapt our activities and expectations appropriately.

5. Community wellbeing

This fifth element stresses that study after study shows that we need to contribute. Teachers, nurses and safety professionals get an element of this from a job that is vocational, but many will get balance from voluntary work with charities and youth clubs. Being actively involved in a safety process that helps keep colleagues safe also helps fill this need. The nice lady who just this morning sold one of the authors some shutters did a professional enough job, but only really came alive when she told the author about the mentoring work she does in prisons.

"It's just dawned on me I've spent half my life
earning money I don't need to spend on things I
don't really want to impress people I don't even like ... "

Even though she was enthusiastic enough about her product not to put the author off buying shutters, it was like talking to an entirely different person when she got on to talking about the voluntary work that clearly inspired her.

More and more organisations now allow workers a period of paid leave during which they can work with voluntary service organisations (VSOs). Indeed, as Generation Y come through, it's clear that the high-fliers who can choose are now using the quality of these schemes to discriminate between potential employers.

Interconnectedness of the five factors

A recent study by MIND found that of the 1 in 10 employees currently struggling with mental health issues around a quarter said that it was because of problems due to work and half because of a combination of problems at work and outside.

You may have heard the saying that a butterfly flaps its wings in a rainforest and shares drop in value in Frankfurt. You understand what they mean, even if you think it unlikely, but everything is interlinked, and it's worth considering some rather more practical examples.

A study at King's College London found that when people sleep poorly it affects the body's ability to regulate the production of hormones such as ghrelin and leptin, which control the feelings of hunger and fullness, respectively. Not surprisingly, they found poor sleepers consume an average of 385 extra calories the following day. It's worth considering the effects of an extra 35,000 calories on the ideal 'beach body' in a three-month period leading up to a holiday. (Over a two-year period, that's more than 280,000 calories.)

Another example: if you suffer from a bad back but hugely enjoy your work, then it's likely your colleagues don't even know about your back problems unless they find you doing stretching exercises in the corridor. However, if you don't enjoy your work, then it's likely to be a very different matter.

From HSE safety research, we know that a 'just culture' approach to incidents correlates with 'above the line' or 'discretionary' behaviour and a decrease in such things as turnover, absenteeism, 'presenteeism' and spurious insurance claims. Perhaps the best-known model of a strong safety culture is the 'Bradley Curve', with its concept of interdependency, trust and 'brother's keeper' behaviour. This equates to the many wellbeing experts who consider the concept of 'care' as vital. This book argues that a holistic approach to individual worker happiness is entirely congruent with sustainable success generally, with a thriving safety culture contributing to this long-term success both directly and indirectly.

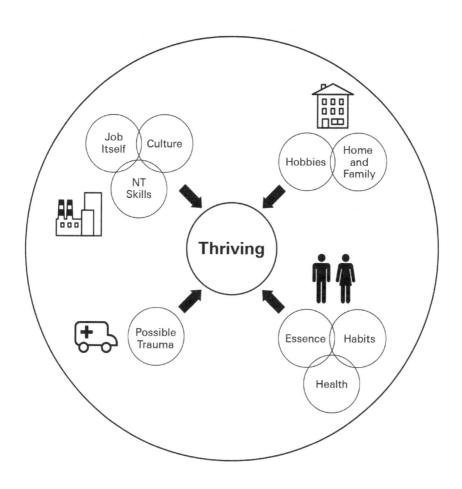

What study after study shows is that, whether causational, correlational or a bit of both, everything seems to cluster. For example, people who exercise regularly are around three times less likely to show symptoms of depression. It's not as simple as 'everything is going well' or 'everything is going badly', but there is a huge link. One of the practical problems this presents is that if you open up a gym on-site, you may be pleased to see that it quickly fills up – but it'll probably be full of staff already healthy and thriving, thank you very much. Those who are already thinking and behaving in a positive way are quick to get involved, but those in most need are likely to be more reticent.

Positive psychology: Seligman's PERMA model

Another well-known holistic approach is the PERMA model, as described in Martin Seligman's *Flourish*. It comes at the issue from a slightly less practical viewpoint but overlaps hugely. PERMA stands for:

- Positive emotions.
- Engagement (i.e. 'flow', where time passes quickly).
- Relationships (family, friends or colleagues).
- Meaning.
- Achievement or accomplishments.

For example, a hard-bitten businessman could lose himself in the act of the deal (engagement) and might derive a great deal of satisfaction from the achievement of stacking up a huge amount of money for its own sake. He may never be happy, though, and may find rewarding relationships in short supply. Similarly, someone might be positive in the extreme but with little meaning. An aimless 'happy hippy', if you would, although we're sure they'd point out their contentment is meaning enough, thank you.

What studies show is that engagement with work gives meaning, and this correlates well with productivity, but that engagement and positivity correlate twice as well with productivity. This means, ideally, with our eye on the bottom line, we need to address the job (it's meaningful and interesting), the person (who feels resilient and positive) and the culture (so does everyone else around me).

Taking a more personal approach, the Austrian psychotherapist who was a founder of the second Viennese school of psychotherapy, Viktor Frankl, author of the multimillion-selling *The Meaning of Life*, states that he found the will to survive Auschwitz so that he could rewrite a book just completed before he was interned. He did survive, and his writings have enriched the lives of millions, as he hoped they would.

A story that sums up Frankl's approach is that of a patient who, after 40 years of marriage, saw no point in carrying on after the death of his wife. Frankl asked if the man would have preferred to go first to be spared the pain he was experiencing. Has was told, 'Of course not', he'd never want his much-loved wife to be going through what he was. Frankl suggested that this was the man's last act of love for his wife. That he survived to cherish her memory, but the price for that was to be the one who bore the burden of loss.

While we're in such deep waters, an interesting statistic: childless adults tend to report themselves to be 'happier' than parents. They have more money, more freedom of choice, fewer responsibilities and constraints, less worry . . . even more sleep, unless they take up the option to carry on partying! But, because of meaning, very few parents would swap with them. There is always a yin and yang.

To summarise 'wellbeing' into a simple five-factor model is utterly impossible. It's too complex, nuanced, bittersweet, frustrating, contradictory, compromised, dynamic and so on. However, the one suggested here: money (basic needs), health, good relations with others, job satisfaction and giving overlaid/interlinked with the more esoteric concepts of positivity and meaning, is, we hope, something we can use as a basic framework.

Five key lessons from safety

4

When working with executive boards over the years, it's been striking how many professional people, despite vast intelligence, expertise, MBAs and the like, are unfamiliar with the grift principle, lack a basic understanding of the systemic human factors and failings people manifest when doing the daft things they do, and why the phrase 'culture eats strategy for breakfast' is so true.

This short section seeks to summarise that key material.

ABC or 'temptation' analysis

It is very easy to use hindsight and blame after the event and point out that the negative consequence was highly likely, but usually it's highly unlikely. For example, every night, in (nearly) every town and city in the world, people are smoking, drinking too much, driving too fast, breaking health-based resolutions they genuinely meant when they made them, eating too much, taking drugs, having unprotected sex with strangers, having sex (protected or otherwise) that they really shouldn't be having, accelerating rather than breaking when the traffic light turns to amber 30 yards ahead, driving when they think they're under the drink-driving limit but they're not sure because they had a glass of wine more than they intended, or they had lots of beers the night before, and, on reflection, that hours/units calculation isn't looking good.

Every year, tens of millions of people will die prematurely because of cancer, diabetes, HIV, hepatitis or in car accidents, or lose their families and much of their wealth in the law courts. On an individual basis, we often ask, 'What did they expect?' But that's often rather hypocritical.

Oscar Wilde quipped he could 'resist everything except temptation', and Stephen Fry said that what he does when he sees temptation is to 'give in to it straight away to save on the faffing about'. If you find those quotes amusing, then we need to refer you to Aristotle, who said that humour is merely 'common sense speeded up'.

You see, every behaviour we undertake has a trigger, but it also has consequences (ABC, technically known as Antecedents, Behaviour, Consequences). In a blame culture, we can often say that because people knew the risks (they were warned, briefed and/or given a toolbox talk), then more fool the person that undertook the risk. However, studies show that it's not the triggers that determine the behaviour; it's the consequences, and consequences can be three things: soon or delayed, positive or negative, certain or uncertain. What studies and, as above, life show is that where consequences are soon, certain and positive, then people will be tempted to cut a corner and take the risk. Delayed, uncertain and negative consequences can be severe, as above, but they may well be rationalised on any given day, and very often the person in question will get away with the risk, and the reassurance from this increases the chance of a repeat.

But those risks mount up, and sooner or later someone gets him or herself into trouble.

Basically, ABC analysis explains why workers put their hands in moving machinery, why we do any number of things bad for our wellbeing and why we fail to do any number of things that would be good for our wellbeing.

The experience of safety shows that education is but base one. The effective response is to proactively identify where temptation is likely to be an issue and design it out and/or facilitate the right choice as appropriate.

From there, it's just a numbers game, and we need to turn to the 'triangle' principle of luck and probability.

The grift principle

In 2012, Matthew Syed wrote a book called *Bounce*. Malcom Gladwell's book *Outliers* covers similar material and is the origin of the oft-quoted '10,000 hours' to be truly excellent at something figure. When we recommend *Bounce* to people, they nearly always say, 'This is a wonderful book.' We've even had highly successful CEOs announce it as a 'total game changer', but anyone with a safety background is apt to comment, 'It's just the Heinrich principle, isn't it?'

The leader of the positive psychology movement, Martin Seligman, refers to it as 'grift' and says that its importance cannot be overstated. Certainly, we feel every school leaver should have a module on it before being let out into the big wide world, and if you're not familiar with the concept, we hope you'll agree it underpins everything in the book, from why the Dalai Lama looks so cheerful all the time through to why dieting doesn't work.

Falling down the stairs, gravity, luck, human factors and common sense

Gary Player quipped that it was funny how he seemed to get luckier the harder he practised. We always imagined him working on his bunker shots at dusk until his hands bled, but it transpires he was in the gym working out decades before any other golfers did. It's a better metaphor. The grift principle states that we get no guarantees either way, but that we can impact on how much luck we might need.

For example, even Mozart, the poster boy for the so-called notion of a 'born genius', wasn't really one. His father was a strict disciplinarian and one of the world's leading music teachers. Not practising hard really wasn't an option. Yes, he was composing while in short trousers, but most of it was rubbish, and he didn't write anything genuinely good until he was in his early twenties. Apparently, some musical experts even consider him to be a late developer!

Called the 'Heinrich triangle' principle, this has been applied to safety for many years. Here's a simple example based on real data. In the offshore oil industry, the likelihood of falling down the stairs is only about 1 in 100,000, despite the stairs being steep, metal and often wet. Consequently, with luck and a following wind, workers may enjoy

an entire career offshore never holding that handrail and never falling. More than that, the unlikelihood of a single act leading to an accident also gives a manager an excuse for a dose of 'blind eye' syndrome so as to avoid potential scorn and backchat. However, over the years, as many workers have been killed in falls offshore as were killed in the Piper Alpha disaster. Holding the handrail simply is a hugely important safety behaviour.

It works like this: stairs will be used about 1 million times a year on a typical platform, so if nobody holds the handrail, we'll suffer 10 falls a year on average. If 90% hold the handrail, we'll suffer one a year on average. However, if 99% hold the handrail, we'll only see one every 10 years or so.

Because all slips and falls seem avoidable, this is an area where blame easily attaches and best practice is easily ignored. The UK Health and Safety Laboratory

(HSL) gives an excellent example of a fatality at a nightclub, which was investigated in depth only because there was suspicion that the man who died had been pushed down the steps by the doormen following an exchange of words earlier, rather than suffering a 'simple drunken fall'.

Instantly, it was apparent that these steep steps were a deathtrap. They were badly lit, with 'handrails' flush to the wall that were essentially decorative only and impossible to grasp. The steep and shiny steps were of different heights, promoting 'cat paw, air steps'. Indeed, the accident book showed that falls and injuries were frequent and that this wasn't even the first fatality.

It's true that most users of these stairs were drunk at the time of their accidents, but instead of a fatalistic 'bad stuff happens' approach, it's better to deal with this clear combination of risks proactively and at least get the basics right. This principle holds in all aspects of life: music, golf, mindfulness, safety, health and driving included.

Wellbeing

The principle certainly applies to the world of health and wellbeing, too. The Dalai Lama quips that his appearing to be happy all the time might have something to do with him 'practising hard at it every day for 60 years'. There are no exceptions. If you want excellence, you need to graft for it. This often clashes with a board that wants excellence through the targeted application of a 'magic bullet'. We see this in safety and wellbeing through the use of motivational speakers and other face-valid but ineffective (in the medium to long-term) initiatives.

People are just apt to do stuff that may be bad for them. Telling them – even imploring them not to – but then shrugging when they fail to follow good advice isn't as effective as facilitating excellence by proactively predicting when they are likely to do bad stuff and putting in some safeguards that will mitigate the consequences. The massive improvement in fatality rate in Formula One following the Ayrton Senna fatality was achieved by enhancing safety design features – not by getting the drivers to slow down and take fewer risks.

Compliance is discretionary

As we have more conversations about holistic coordinated wellbeing/ empowerment processes we're often hearing the challenge:

This empowerment stuff sounds great . . . but is it not a case of running before we can walk, as we still haven't reached an acceptable level of basic compliance for several of our key processes?

The observation from the world of safety culture that informs the response is the truism that, very often and in many important ways, *compliance itself is discretionary.* Of course, the less direct supervision there is the more this tends to be true, and in many respects, the interplay between autonomy, rules and individual volition underpins everything in this book. Once you have hit diminishing returns with rules and processes, more rules and processes will just spin you in ever decreasing circles. Taking a proactive, human-focused approach is the only way to achieve a step change in outcome. It's a key mindset shift undertaken by every organisation that has achieved safety culture excellence, and it applies at least as much to wellbeing.

We need to tackle something really controversial here. Influential writers like Barbara Ehrenreich (*Smile or Die*) rip into 'positive psychology orthodoxy' and point out that wealthy people often have healthier lifestyles because it's easier for them to do so. Front line jobs are typically more stressful, and people who do them are more likely to drink, misuse drugs (including legal ones), smoke, collapse into the sofa and eat badly. (Yes, that's not entirely a one-way relationship, but it's certainly also ABC analysis 'getting to the end of the day' in action). She also points out that smoking and obesity are increasingly seen as option-limiting 'class-markers'. In short, she's pointing out that polarising, self-perpetuating vicious and virtuous circles abound and suggests society is guilty of blaming and scapegoating. It's a hugely valid point, but if you're an *individual* reading this then another point well worth making is that we all have a choice because almost anybody can apply the grift principle to anything. If you're a heavy smoker who's 4 stone overweight, aim to be a moderate smoker who's 2 stone overweight. And repeat . . .

> The richest 20% of the population ride a bike 5 miles for every 1 mile the poorest 20% ride . . .
>
> (Department of Transport Travel Survey 2018)

Just culture: consistency, transparency and fairness

The truism 'culture eats strategy for breakfast' reflects the fact that the culture is not what we say it is; it's what actually happens in the middle of a busy day.

It's 'the way things are around here', and it manifests through the massive power of norms. In short, no matter what you are told in the induction and training, if the old hands don't do it, then it's highly likely that the subcontractors and new starts won't be doing it by the end of their first week.

Perhaps the most influential theory that has its basis in culture and safety (that isn't temptation analysis) is that of the just culture. When suboptimal actions are analysed objectively, it can often be seen that many are entirely 'unintentional errors', often because the person doesn't realise the risks they're running. You can't blame people who don't know what they're supposed to be doing, or doing the best that they can with the tools and/or time they have available.

The second group of suboptimal actions are called violations, and they are distinguished from errors, as the person knows them to be suboptimal and chooses to do them anyway. Often, though, that's 'chooses'. These break into three categories – situational, cultural and individual.

A situational violation is a variation on the old maxim 'you can have it quick, cheap or good quality – pick any two'. It's the basis of the union terror tactic that is 'Work to rule!' Cultural violations overlap greatly with situational ones and occur where the person is more actively cued to do something suboptimal by the environment. Classically, where the norm is that everyone does X so nobody new wants to stand out by doing Y is where they are told, 'We want this done safely, but by Friday.' (If you've ever been told, 'You're a really nice person and I really like you, but . . .,' your heart will have sunk instantly because we all know that the meat of a sentence follows the 'but' and that the words before that 'but' are so much flannel.)

In short, 'safely, but by Friday' 'tells' you what's expected, and we are hardwired to give our seniors what they expect.

A senior NASA figure commenting, 'I hear the contractor's concerns, but I'm appalled by this no-fly discussion on the eve of a launch' (and just before a major contract renewal) meant that the engineers' severe concerns about the cold weather causing the seals to fail were overruled and that the Challenger was launched only for the seals to fail and the Challenger to explode, as the engineers predicted.

When analysed objectively, what we find is that rarely is the person 'off on a folly of their own' and that in the vast majority of cases even violation behaviour makes sense to them on the day. In safety, we've learnt to ask the question, 'Why are you doing that?' curiously, not aggressively, as it nearly always leads to an answer we can work with. We've also learnt that the proactive question, 'Anything slow, uncomfortable or inconvenient about doing that job safely?' also nearly always leads to analysis and information that allows us to design out

WE HAVE 'NO NAME, NO BLAME' REPORTING ON THIS PLATFORM– IT'S JUST THAT MANAGEMENT REALLY LIKES TO KNOW WHO IT IS THEY'RE NOT BLAMING
ROGER BEALE

a human factor risk. Similarly, the core of the 'safety differently' approach is to proactively ask, 'How can I help you do this job productively and safely?'

Finally, from ABC analysis, we learn that even behaviour that can be classed as 'individual folly' is often eminently predictable and that practical solutions can be implemented to mitigate the damage, like putting up signs to take care by the stairs of a nightclub.

Bullying

Bullying tends to go hand in hand with a blame culture, and vice versa. The more often the organisation is thinking objectively and constructively then the less 'space' there is for bullying. It's not a small problem, as Britain's Healthiest Workplace survey had 1 in 20 saying they were bullied regularly and 15% saying that they had been bullied in the past year.

What's interesting is that not only is bullying less tolerated in a 'just culture' but we are more likely to see more proactive, constructive thinking and behaviour around the subject. For example, the more progressive programmes will counsel not just the people being bullied but, mindful that most bullies are working through psychological issues of their own, will also seek to counsel the bully before or as well as any punitive action.

Organisations are rarely as good as they think they are

When formally assessing company culture it's of course important to compare management scores with workforce scores. Over the years, we're come to assess whether the gap between the two is what we came to label the 'natural gap' of management overconfidence or something more concerning. (Arrogance? Denial? Projection? A blame culture?) In the absence of these, the comparison often triggers a useful and constructive debate about perception.

MIND recently found that whilst three quarters of managers feel they are supportive of mental health issues only half of their employees agree with that view. There's a very good chance an organisation will read through some of the lists of best practice listed in later chapters and say to themselves 'broadly speaking, we do all that'. Even then, it's a very good idea to systematically and robustly check with the workforce, as the data from MIND suggest that one in three companies are in for a nasty shock.

Why applying a safety programme based on these five principles should be a key part of a holistic wellbeing programme

Applying the principles in the first section will mean that traumatic harm and witnessing traumatic harm will be far less likely. When incidents are run through decision trees pre-agreed with and part implemented by the workforce, it increases transparency, consistency and fairness. Fairness is the number one societal value and is judged far more harshly than illegality even – laws just being a way of trying to ensure people behave fairly most of the time.

When meta-analysis is undertaken of workforce surveys, one core factor tends to stand out as dominating the workforce perception of management and the behaviours that flow from that unspoken psychological contract. It's: Does the workforce trust them? Do they trust that management are doing everything reasonably possible to keep them from harm while they are making money for them? Do they trust management to treat them fairly if something goes wrong?

The first element means more than just telling employees what they need to do to look after their wellbeing; that's just base one – it's systematically using proactive techniques to help them do that. The next chapter looks at mental and physical resilience, then we'll look at minimising the impact of the number one cause of workplace stress (the soft skills deficiencies of your boss). Finally, we'll look at making the job itself more rewarding and satisfying (so you'll need less resilience to deal with it!).

Individual resilience

5

One of the godfathers of organisational resilience, James Reason, has some Golden Rules of Error Management. These can be boiled down to:

- Error is inevitable (people are involved and we learn by trial and error), and even the best people make mistakes.
- Error comes in patterns.
- We can't change human nature (but we can proactively look for those patterns and change the environment).

And

- Error management is about making Good People Excellent (and, to speak bluntly, crap people adequate!)

We feel this is an excellent reminder that an objective learning focus is key and that whilst we must always ask ourselves not 'is this person fit for work?' but instead 'is this work fit for our people?' empowering and energising individuals will always be part of a holistic approach. So, whilst 'resilience should be last, if at all' is a good maxim for reminding us to avoid putting the onus on the individual when planning a strategy, from a book perspective it's a good place to start before broadening out into organisational issues.

And there are times when individual resilience is most certainly required.

Studies show that whilst many people don't change much – people can (and do) change much more than we used to think. We just need to physically build and/or strengthen positive neural pathways in very much the same way as we build muscles – by targeted repetition. This is very useful, as we 'need a new version of ourselves for every new stage in life' as the saying goes . . . or if you like: it may have been crap up until now but it's never too late!

Firefighters, paramedics, police officers, social workers and others have to deal with things that would drive many of us to sleepless nights. But it's not just the 'blues and twos' and those that interact with them. Reporters working on child abuse cases face trauma. Human resource staff involved in 'rightsizing' exercises, who aren't hard of heart, face trauma. Cold-callers and call centre staff told to 'eff off' 50 times a day don't have it easy either. Nor do teachers in difficult schools, anyone working to a challenging deadline and anyone who hasn't been trained properly and/or hasn't the correct tools or skills. Then, of course, there are those that hate their direct supervisor because they're horrible to them or who are being bullied.

Then of course we need to recognise that daily life is really much more stressful than it used to be. We get cold-called and bombarded by pop-up advertisements to distraction. Checking in at an airport or ringing an organisation to speak to someone about a minor technical problem can be utterly infuriating. Buying a ticket for a concert or opening a bank account are both 100 times more bureaucratic and time consuming than they used to be. Or so it feels. The old *Little Britain* joke 'computer says no' *just really isn't very funny anymore!* 'Dalai Lama in an expletive-riddled tirade and meltdown' is a headline just waiting to happen.

The list goes on and on, including the 75% of workers who say that they are disengaged at work because finding work boring is stressful! With many of the 25% who do enjoy their work, despite the challenges, as above, it's actually quite difficult finding anyone who wouldn't benefit from working on their resilience!

The following material considers first the resilient mindset and then what is becoming known as 'alertness creation'. This reflects the basic principle of intelligence, which, in essence, boils down to the fact that we have a reptile or instinctive brain and also a more recently acquired thinking brain. The fewer instinctive mistakes we make by habit and the more constructive and active thinking we do, the better. Reading is a good example. If we teach ourselves to read quickly and accurately, we misunderstand less and have more time and energy to process the information.

Excellent but tired readers can find themselves 'skating' over the text, taking in nothing, but even well-rested readers will make mistakes. For example: assuming they know exactly what a word means when actually they, like most other people, don't and the writer is one of the few to use it correctly. (We're thinking about the likes of Stephen Fry or Ian Hislop!) We'd give a personal example or two, but we don't actually know which words we only think we know the meaning of . . .! A famous safety-related example: at Le Mans in 1955 a terrible accident killed 84 people when a car flew into the crowd. The fatalities could well have included the reigning world champion, Juan Fangio, but, with seconds to spare, he instinctively slowed down as he rounded a blind corner at 200 mph. When asked how he could possibly know something serious had just happened ahead he explained that he was the world champion, driving a silver Mercedes at more than 200 mph but that no one was looking at him. This wasn't ego at play but a dramatic example of excellent situational awareness.

Resilience enhancement is about ingraining good habits that allow us the time, space and energy to do more constructive things and think more constructive thoughts. As above, both vicious and virtuous circles abound with both front and back brain contributing simultaneously. For example, when presented with an offer we consider unfair, brain imaging shows that, automatically, the parts of our brain that light up are to do with conflict and disgust. However, when presented with a fair and reasonable offer the parts of our brain that automatically light up are to do with planning and cooperation. This is the well-known concept of the 'psychological contract' in physiological action.

A lovely example of the need to engage the prefrontal cortex in some active thinking comes from the footballer Tony Cascarino. He played football on the afternoon his first child was due (it was a different era), but on hearing he'd just had a son, he rushed to the hospital straight from the game, signing autographs as he went and stopping only to buy his wife a card. But when his wife read it, she got very angry and threw it at him . . . in autopilot, he'd signed it 'Best wishes, Tony Cascarino'.

Of course, these autopilot cock-ups never happen in an occupational setting.

The resilient mindset

We all know that resilient people can be big or small, strong or (physically) weak, so it's worth stressing that resilience is about mindset – a state of mind –

and it is a hugely essential part of the interconnected factors discussed throughout the book.

For example, in *Flourish*, Seligman quotes a major US study that shows stress is, of course, bad for you. Unless you hold the belief that it isn't, in which case it isn't harmful! We're oversimplifying, but studies show that it's definitely harmful to people who think it is and a bit harmful to those who are more middle-of-the-road. But for those people who can take life's slings and arrows head-on it's not a problem.

Hygge

No book on wellbeing written in 2017 could fail to mention hygge, the 'secret of Danish happiness'. There must be more than 100 books on the market by now, and this short section could of course feature anywhere in this book, but we've included it here because of an article that discusses its root cause.

For the readers who haven't come across this concept yet, recent studies have suggested that the Danes are the world's happiest people, and they put this down to the hygge mentality. This is basically the habit of getting together in a room warmed by a real fire, wrapping themselves in furs and warm blankets and drinking wine and eating chocolate . . . in moderation, of course. The very definition of 'a little of what you fancy does you good'.

The article in question, in *Le Temps* magazine, postulated that when the borders of Denmark were shrunk in the nineteenth century, the nation effectively collectively decided that instead of stressing about lost power and status they'd 'embrace it, 'identify with simplicity', chill out and embrace the quiet life. Military defeats by the likes of Germany, Iceland and Norway were not bemoaned as losses but celebrated or, in modern parlance, 'reframed' as gains. Similarly, the Dutch, who also had pretty rough nineteenth and twentieth centuries, but who also model the very definition of 'chilled out', are said to ascribe their happiness to 'biscuits and hot chocolate while raincoats dry in the hall'. We'll never trace where that came from. (A *Sunday Times* writer on a TV panel – Rod Liddle – quoted the Dutch as saying it. But it doesn't matter. We all know exactly what they mean!)

You may have read one of the world's best-ever-selling management books, *Who Moved My Cheese?* It's the simple parable of two mice – one who notices that the cheese store is depleting and looks for alternative supplies and the mouse that, in denial, does nothing until the cheese actually runs out. You don't need us to tell you how it works out for the mice in question!

The simple message is that change is inevitable, and those who embrace it will probably flourish, and those that fight against it will probably struggle.

A very practical application: post-traumatic stress disorder and post-traumatic growth

There's a famous expression that 'What doesn't kill you makes you stronger'. Obviously, there are times when that's just nonsense, but there's truth in it too. The worst experiences, even torture and bereavement, can be turned to a positive if they increase self-confidence, perspective, humility and self-knowledge. It is true that with the right mindset we can learn from almost anything.

Nelson Mandela is perhaps the most famous example. When locked up on Robben Island, Mandela would make a point of spending as much time as possible talking to his guards. Not just because he was trying to convert South Africa one person at a time, though he was, of course, but because:

> In genuinely listening to them, I learned so much about the Boer mindset. Their values, hopes and fears. It stood me in such good stead later when we opened formal negotiations. I had more understanding and more respect.

Anyone who has seen the film *Invictus* will know that Mandela's suggestion that the national rugby team keep their jersey and anthem went down like a lead balloon with the ANC and that he only took the vote by a narrow margin, even though making it a resigning matter – with this in 1994, when his moral authority was huge. As the film makes clear, he was right, and the whole country united behind the team, and it was one of the best instances of sport being used to unify and heal old wounds.

A second point needs to be made about transformational leadership. We know that you can influence a person by the very act of listening, and we all know just how many of these guards later attended his presidential inauguration, shedding tears of joy.

Kintsugi vases

A useful concept that all who come across it seem to love is that of the Kintsugi vase. The Japanese tradition is to repair jugs, vases, etc. if they become broken and smashed, but rather than try to hide the cracks, they are augmented, even celebrated, with gold leaf. The really rather delightful notion is that the piece can actually be more beautiful and is certainly more unique for having been broken and repaired.

Would you bet on a horse ridden by a jockey who had fallen off hundreds of times, broken every bone in his body and lost more races – 14,000 plus – than anyone else ever? You would if it were Sir A.P. McCoy, *also* the rider of more winners than anyone else ever, with more than 4,000.

Some other ideas that might form the basis of a workshop

Collated from a variety of books and sources and not necessarily in order:

1. Be here now

Every second spent wishing what's happened hadn't happened is wasted. You can only plan what you're going to do to make sure it doesn't happen again – or get on with doing it. A variation is pinning your happiness on the future. This is the classic 'I'll be happy when . . .' mindset ('When . . . I get that job . . . win that money . . . pass these exams . . . buy that new phone . . . get promoted . . . I'm on that holiday . . .'). Of course, when we get there, the actual experience is often not at all what we hoped for or expected, and we can feel crushed and disappointed.

Consider Newport 'super-tramp' poet W.H. Davies' most famous poem:

> What is this life if, full of care,
> We have no time to stand and stare.
> No time to see, in broad daylight,
> Streams full of stars, like skies at night.
> No time to turn at Beauty's glance,
> And watch her feet, how they can dance.
> No time to wait till her mouth can
> Enrich that smile her eyes began.
> A poor life this if, full of care,
> We have no time to stand and stare.

In a similar vein, Einstein quipped that anyone who can kiss and drive at the same time just isn't giving the kiss the attention it deserves! Mindful eating of chocolate is a bit of a gimmick, but it's true that if you unwrap it slowly, look at it, smell it, relish the prospect of eating it and then strive to savour every small bite, you're likely to take on less calories than if you just grab some and mindlessly stuff it in while doing something else!

One of the authors handed small bars of chocolate out at the start of a conference talk to do the gimmicky mindful eating routine. Then, 10 minutes into the talk, the author tried to start the exercise – but having forgotten to say 'don't eat it yet', there was no exercise!

This is, of course, where the section on mindful meditation or just basic mindfulness goes. It's ubiquitous (there are now several mindfulness apps!) and perhaps far too trendy at present, but it's also scientifically proven to be, beyond doubt, hugely useful in achieving just about anything of importance in this book: physical and mental health, concentration, success, happiness, successful relationships and the ability to deal with pain and setbacks.

So if you haven't yet tried it, you probably should. It might not work for you, but it probably will. Books are plentiful, and there is almost certainly a selection of classes in a hall near you.

2. Be humble

We can get into the habit of trying to be right all the time, and many people who get some 360-degree feedback find that others have them on a continuum from 'thinks they're right all the time' to 'knows they're right all the time'.

But we're not right all the time, and we should, all of us, be more humble and listen more. This is easier to say than do, yes, but as well as relationships it generally plugs directly into the empowering methodologies discussed in this book.

In his books on team excellence, Leoncini stresses that 'being humble' is one of the three great virtues (along with focusing on the team's success and people skills). He stresses that being humble doesn't mean being subservient, nor does it mean not standing up for yourself when treated unfairly. Rather, it means being low in ego.

For example, not saying something that needs to be said for the good of the team because it might make them feel uncomfortable is a lack of humility in this context, as it reflects a mindset that's too much about 'me'.

3. Write your own (ideal) obituary

When you're done with this life, what do you want it to say? Are you working towards that, in which case your life will most probably have traction and meaning, or have you gone off on a tangent, in which case you'll probably be suffering from existential angst? To quote the storyline from the Oscar-winning film of 2017 *La La Land*, you can take a job you don't like much while saving up to buy a jazz club, but once you've saved up you have to buy it.

One of the authors was tasked with some stress management some 20 years ago with the high-flying graduates of a pharmaceutical company. They did this exercise on day one. Their very best graduate (double first from Cambridge, rowing blue and a delightful woman) didn't turn up for day two. She was on her way to Africa to do voluntary work. The head of HR really wasn't very impressed at all – but this is the bit of the book about individual resilience.

4. Be grateful (un)dead

We are all surrounded by 'miracle and wonder', to quote Paul Simon. Every day, you'll have crap to deal with, of course, but every day is also full of wondrous riches. The smile of a child, a sunset, a nice cuddle, just the memory of a smile, a good book, laughter, fresh air, the memory of Wales beating England at rugby, the prospect they might do so again one day . . .

Last year, one of the authors took his son to see the Grand Canyon. At the hotel, they met a woman in a headscarf coming to work who was clearly ill with cancer. Asking after her wellbeing (she had stumbled), she assured the author she was fine and explained she had no healthcare so had to keep working, and added with a big smile, 'But I ain't complaining . . . the sun is shining and I'm well ahead of the game . . . well, today at least!'

Spending five minutes a day actively being grateful for a handful of things really helps a positive mindset and perspective, and there are construction companies encouraging workers to keep gratitude diaries.

5. The balanced but

As above, one of the things we know from safety is that the word 'but' can be a killer, figuratively and literally. Saying that we want a job doing 'safely, but by Friday' means by Friday, and everyone is hardwired to give management not what they claim to want but what they really want. Though afterwards the claim might be made, especially in a law court, 'I said safely, I did say safely', the damage is done on the day. It's just like when a partner says across a table on a date, 'You're a really nice bloke and I really like you, but . . .' That sentence never ends well.

In short, we all know, subconsciously or otherwise, that the most important part of the sentence is after the 'but', and it's true of thinking or cognition too. 'My manager has just given me a terrible briefing, and I'm not sure what to do . . . what an idiot' is perfectly fair and reasonable. 'That was a truly terrible, not to mention dangerous, briefing, but they're usually supportive and I know she's under time pressure and really stressed, so I'd better seek clarification' is rather more constructive.

Virtually all things come with positives and negatives, so training yourself to order your cognitions negatively but also positively, at least when it's viable and/or fair to do so, can become a life-changing habit.

6. Every breath you take

When someone has a panic attack, the first thing a psychologist will get them to do is slow down their breathing. Deep and slow is ideal to regain control. The same is broadly true in everyday life and relates to our sympathetic and parasympathetic nervous systems and a thing called the vagus nerve. The sympathetic system is 'fight or flight' and relates to dealing with threats. The parasympathetic system is 'rest and digest', where our vagus nerve gets stroked, like a cat if you will, and we sort of purr away.

In short, we need to be parasympathetic every few hours or so to maintain balance, and everything in this book helps with that generally. Breathing correctly (i.e. deep and slow) really helps with that at any given moment. Here's an exercise you can do while you read. Take some deep, slow breaths in then out. Then, at the end of a deep breath out, 'double-puff' and force an extra bit of exhale from deep in the lungs. You may feel a bit weird and tingly but you're probably also feeling pleasantly calm!

One of the authors used to counsel nervous students before exams, telling them that if they do nothing but breathe deeply when they turn over the paper it'll ensure they can think clearly and not panic. To focus them, students were asked to score themselves on their breathing in the first few minutes. No student who reported a score of 8 or above on a 10-point scale ever reported a problem with panic and focus.

7. Commit to taking control of the dice

We discussed above the lessons of the grift principle and that though we get no guarantees either way, we by and large get the luck we deserve. It's a universal truth that the more often we 'own it' and 'work it', the better.

8. Achievement and failure go hand in hand

Accept that, like Tony McCoy, failure goes hand in hand with success, especially for those who strive to excel or innovate. We cannot avoid failure, because much learning is based on trial and error, but we can learn as we go. As the saying goes . . .

You cannot fall if you do not fly. . . . but what is a life without flight?

Summary

Buddhist philosophy can perhaps be boiled down to the following: that life is full of wonder and miracle but is finite (so best make the most of it) and that good and bad will happen to all (starting with the fact that everyone must die). The only thing we can control is how much we relish the good and deal with the bad – especially as these will influence those merely watching us, as well as those we interact with.

We don't know if Martin Seligman and others quoted and summarised here are Buddhist, but it doesn't matter. The point is that there's clearly a lot of consensus about this mindset material, and there has been for quite a long time! Buddhists would be the very last people to claim they are the last word on this mindset material, but no one ever said 'What are you, some sort of Buddhist?!' as an insult.

Knowing yourself and your impact on others

The first step on most resilience courses is to understand yourself and to understand your impact on those around you. There are a million and one different personality questionnaires, but the better ones are always some variation on the 'big five' personality factors that studies show cluster. In plain English, these are:

1   Extrovert/introvert. Extroverts tend to be friendly, outgoing and energetic, but also impetuous.
2   Controlled/anxious. We all suffer 'fight or flight' impulses every day. How well do we deal with these?
3   Conscientious. Some people are diligent and detail-conscious; some are rather more relaxed about such matters.

4   Agreeable. Some people are easy to be around, empathic, humble, straight-talking – basically bulls>- and bluster-free. Others are not!
5   Curious and open-minded. Some people have a learning focus and don't just want to understand but need to understand. Others were born already knowing everything or just not interested.

A suggestion about the use of personality questionnaires: the first thing to note is that most are self-report – so it's hardly surprising that most delegates say, 'Yes, that sounds like me' during feedback and coaching sessions. (It works like this: Q1: 'Are you an extrovert?' 'Yes I am!' Then later, 'The report I have here says you're an extrovert' 'Hey, what a good report it is – it's got me in one!'). The use of 360 feedback helps overcome this problem – where colleagues, reports and supervisors are asked for their opinion. We're nearly all deluded about ourselves to some extent – with this ranging from a little bit under-confident through to massively overconfident. It's good advice to only use quality personality questionnaires at any time, but this is especially true when using 360 feedback. Well-constructed question sets and normed scores and scales help when such profiles are given back to employees.

Belbin's Team Role Questionnaire is freely available, and a discussion of the 'strengths and weakness' of each preferred role can often provide enough of an insight to get employees reflecting as we'd wish. For example, the 'monitor evaluator' role is often essential for a successful team, as they naturally take a 'helicopter' view and are usually right. However, if that evaluator hasn't the soft skills to point out that the team has gone off on a tangent in a way that doesn't alienate everyone, then that natural tendency to be correct may not be much use! They may not disrupt the group for long, though, as evaluators are more prone than most to sulking if the team disagree with them. Typically, after the third 'You useless lot have gone wrong again . . . why don't you listen to me?!' has been ignored, they'll simply withdraw into themselves or physically walk off.

For most of the group, that'll be welcome. But not for the natural 'team player', who doesn't like to see anyone upset and will continue to try to pull them back in, but because team players are usually modest and self-effacing, their vital but often understated role can be underappreciated. That is, until they leave for some reason and the team promptly falls apart.

You'll recognise the dynamic described above, we're sure, and you'll know that a feedback exercise about 'your typical style and its impact on others', done well, can really help individuals achieve their maximum potential for themselves and their companies. You may well have experienced this sort of feedback yourself.

A warning: it's all about the quality of the reflection and the skill of the facilitator. Just buying the most expensive personality questionnaire will prove little use to an organisation if it's not used well.

The importance of striving for objectivity

The concept of 'own it' and 'use it' because everything in your life comes with one common denominator does come with a caveat, which is the objectivity of your introspection.

It's important not to blame yourself when things go wrong if actually it's the environment that is the major cause, though far more likely we'll conclude 'objectively' that it's the environment that was at fault when actually it was us. (The typical person is overly optimistic about their abilities. For example, there is a classic study of driver mentality that showed that 48% of drivers think they are above average in ability and 48% that they are average. Worse, most of the 4% 'below average' said that the problem they were confessing to is that of overconfidence because of their wonderful hand-eye coordination and superb reflexes . . .)

We all have our biases and blind spots, but central to minimising blame in a culture is the concept of the fundamental attribution error, which shows that when it's other people who have done something suboptimal, we tend to blame them and underplay the environmental causes. When it's us, we can list all of the environmental causes and quite possibly some rationalised excuses too.

There are many, many mistakes we can make and thousands of books and courses to reference, but to summarise a couple of main ones:

- We can all be prone to mind-read and assume the person meant something when actually they didn't, and we've inferred meaning and/or intention that wasn't there or underplayed the role of the environment so that we take something too personally. (A golden rule in mental health work is 'don't take it personally'.)
- We 'reason emotionally'. People who are prone to emotional reasoning are the sort of people who consider that the proof that it's true is that they feel strongly that it's true. This is circular reasoning, and they may reframe, deny, distort and rationalise contradictory facts almost indefinitely. It's very difficult to engage in productive dialogue with people with a ready supply of 'alternative facts' or who state 'we've had enough of experts'.

This subsection includes the words to strive to because: (a) any philosopher will tell you there is no such thing as objective truth; and (b) clearly these things apply to everyone. We'd just like to add that: (c) some people are making a better fist of minimising the hit than others!

Joking aside, we all suffer from 'cognitive dissonance' to an extent where how we behave and the person we claim to be diverges. Dealing with it on a day-to-day basis can be tiring and stressful, and being presented with proof that we've been deluding ourselves can be hugely traumatic. It's worth making the effort to minimise the dissonance.

Some interesting research

It's long been accepted that few people change personality significantly, but recent research suggests that's not particularly true and that actually the average person does change as they get older. This is both a greater opportunity and a threat than first thought, as what becomes key is in which direction they change!

By the ongoing repetition of new behaviours, we can establish habits (or physically, faster and stronger neural pathways that reduce the 'effort' required to undertake these new thoughts). For example, people who regularly meditate, for example, do become significantly calmer and more agreeable. (This can be physically measured – for example, they jump less high when something goes bang!) Highly introverted people can become highly effective, engaging and interactive presenters through practice and experience. (The writer Marcus Buckingham jokes that while he can comfortably present to an audience of thousands, he is, however, a real introvert and confesses he still finds one-to-one small talk at the aftershow reception exhausting.)

Some people will find soft skills harder than others to master – but everyone can achieve a suitable level of mastery. It just takes practice and repetition until it becomes habit. The most important factor, though, is locus of control. Some people are naturally high in it; others naturally low. Above, we covered the grift principle, which stresses that, on balance, we get the luck that we deserve, and though some may be naturally more industrious and conscientious than others, we can all learn to 'own it' and 'use it' more and be less fatalistic.

Energy and alertness creation

This used to be called 'stress management' but that typically involved a reactive response to people who were struggling. A proactive approach to all is far more effective. Experts suggest that it all revolves around sleep and that it's a circular relationship. If you are sleeping well, you are probably alert and energetic and know that you are preparing for a good night's sleep from the moment you wake and throughout the energetic and alertness-filled day to come.

> People who sleep badly consume around an extra 300 calories the next day in an effort to make up the energy gap . . . not always by way of apples and raw carrot sticks . . .

Sleep

With advances in science with brain imaging, it's only in the last five years or so that the importance of sleep to wellbeing has begun to be fully understood. Previously, what we could see (rapid eye movements, or REM) was considered important, but it turns out that memories are stored and tired neural pathways reinvigorated during deep sleep (essentially, when the body feels we are really safe, and so attention can be devoted to such maintenance tasks). We all need this deep sleep or we will suffer sleep deprivation and our cognitive function will be impaired. This means operating on autopilot even more often than usual, which of course means making mistakes and impaired situational awareness. It also means it's harder to find the energy for those proactive things that require effort (such as coaching).

If we have super-efficient brain maintenance, we can sleep for only three or four hours and awake fully recharged, but the typical person requires six hours or more. Einstein apparently was one of those who could function at 100% on around three hours. We'd like to think that we too could have predicted with nothing but pen and paper that imploding black holes could warp the space-time continuum, given that extra 35 hours a week to work on it, but then again, perhaps not! (The well-known 'eight hours' is, apparently, a myth invented to protect workers being exploited by employers, who'd have them working 18 hours a day if they could).

A simple example: we know we should strive to 'drive to the distance' to maximise anticipation and reaction time and switch on fully while undertaking the highly dangerous task that is driving, but sometimes we simply tail the car in front and arrive home with almost no memory of the journey. This is because though only 2% of our body weight, our brains typically take up 20% of our energy, so it'll of course switch to energy-saving mode whenever it can. This can include microsleeps while we are driving, and the data from recent studies as to the frequency of such events are truly frightening.

"We need to revisit the shift rota and run some alertness creation workshops ... or we could just remake Zombie Apocalypse ..."

Not everyone knows that we can only adjust our body clock an hour a day (no matter what formal international regulations say is safe), or that once you've been awake for 18 hours your cognitive function is on a par with someone several times the legal drink-driving limit. Indeed, the physiology of drunkenness and tiredness is almost identical: clumsy, poor hand-eye coordination, lack of concentration, emotional instability.

One of the authors once ran a session for a client where the topic of drivers having 'silly' incidents came up. In many of these cases, their 'excuse', it was reported, was that they were tired. When we dug into the statistics, it was clear that this correlated directly with shift changes. The CEO stressed that he'd confirmed that their shift system met EU guidelines. We put those aside and had a productive session that started with a discussion of the above physiology.

Causes of poor sleep

You don't need us to tell you that crying babies, relationship stresses at home, a party lifestyle, physical and mental health issues and medication (self or via the pharmacy) can impact on good sleep. Work can as well, of course. Shift systems are an obvious culprit, but monotony, too little variety or too little role clarity, and too little autonomy (or too much), can all also cause stress. A culture of urgency and high achievement can, of course, prove stressful/lead to poor sleep.

One of the world's leading experts in the field, Dr Susan Koen, concludes that as we understand more about the way the organisation, the task and the person interact, we need 'difficult discussions and an honest appraisal about how we use the human resource'.

She suggests the essence of the high-reliability organisation is the model we need to follow. And since alertness is at the epicentre of the human factor issue, all companies should follow it:

- Identify the risk.
- Minimise its opportunity to cause harm with design and monitoring processes and procedures.
- Ensure we can quickly identify and quickly and effectively respond when the risk factor slips through the systems we have set up to contain.

This might involve proactively asking people how alert they feel to identify problem areas to be addressed, or nap rooms that employees can go to when they feel really tired. A 20-minute nap hardly makes up for a major sleep deficit, but it does give around two hours of alertness, which might be vital if, in that two hours, there's some heavy machinery to be operated, a risky task to be undertaken or a critical meeting or phone call to make.

Sleeping well

Some things not to do:

- Stress-bust with large amounts of alcohol rather than exercise and/or meditation.
- Drink caffeine after 2 p.m. (especially in older people, who don't sleep as well anyway and become less efficient at breaking down and absorbing the active ingredients).
- Exercise close to bedtime to 'tire ourselves out'.
- Have electronic devices in the bedroom to use just before we sleep (which stimulates the brain) and that glow in the dark.
- Keep irregular hours.

As well as avoiding the above, we'll have regular hours and retire at a similar time each night, with a milky drink – and we'll retire to a cool bedroom ('open window, warm duvet' is the expression often used).

Dealing with that mid-afternoon energy dip

If your company has a treadmill workstation, then take that list of phone calls you need to make and spend the next 40 minutes 'walking and talking'. If the weather's good enough, then why not make those calls while walking around the industrial estate ornamental lake in the sunshine? It's said that men especially communicate really well walking alongside each other (for ancient historical reasons to do with hunting). Just because that walk around the ornamental lake in the sunshine is extremely pleasant doesn't mean a handful of employees can't knock off a really astute SWOT analysis of a key issue then return to their desks refreshed. With Land Securities based near the charming St James's Park in London, many a walking meeting is now held in it.

There's far less leeway with production line work, of course, but even so, there's no legal requirement to hold a production meeting or a toolbox talk in a dank, airless office.

Finally, advice from a £5,000-a-day 'executive energy coach' seen on a TV couch: if there's nothing else for it, he suggests, then find a quiet spot and knock off 30 quick press-ups or squat thrusts/star jumps – you get the idea.

A safety-related case study: Buncefield

At Buncefield, many operators were described as shocked, bemused and struggling to understand what had happened, saying, 'That simply can't be the case' and, 'I just don't understand how that happened'. With workers undertaking a punishing switch shift programme, as well as averaging huge amounts of overtime as standard, confusion was entirely predictable.

- Workers were doing three-day 12-hour shifts then switching to four days of 12-hour night shifts.
- Overtime could be up to 120 hours a month and averaged 50–60 hours!
- Workers were eating meals at their desks because of pressures of work.
- Witnesses were afterwards confused and bemused and (genuinely) saying, 'I just can't have got that wrong' and, 'I just can't see what went wrong there', as well as other signs of being oblivious or being in genuine denial (both strongly suggesting a confused and/or fatigued mind).

The mistakes at Buncefield helped cause the biggest explosion in Europe since the Second World War, but we can all make smaller, less spectacular but personally or professionally catastrophic mistakes in similar circumstances.

There's no point working more than 40 hours

It's worth reiterating the issue of long hours and productivity, as research suggests that after a certain amount of time, productivity flatlines or even drops back. In short, long hours add nothing to the organisation except added risk. You may be surprised to know that the definition of 'long hours' is 40, so an excellent question for a CFO is 'Why are you paying people 10 hours overtime a week to achieve the same amount of output as someone doing the standard 40?'

Fuelling up: caffeine and sugar

This topic, of course, edges us towards the old joke with the punchline, 'If I do these things, will I live longer, doctor?' 'Maybe a little, but it'll definitely feel like it.'

Constantly fuelling up on coffee and sugar is bad for you. We all know this. Ideally, we'll get all the hydration we need from water and blended vegetables and all our sugar from eating fruit – though not too many of those apples, which have been bred to be so sweet they now come with 'medium sugar' content on the label. For many of us, this zero tolerance approach simply isn't a starter! Trying to pass a kebab shop when walking home full of drink is the very definition of an ABC analysis temptation. Talk about soon, certain and positive! (Incidentally, the first time one of the authors checked how many calories there actually are in a bottle of wine, they cried!)

As well as walking and exercise, as above, the practical steps we can take are to remember the grift and hygge principles and just cut down. Some top tips:

- Eat food mindfully and slowly. Chew it longer to break it down more with saliva (as it's supposed to be) and savour the taste – don't just wolf it down mindlessly.
- When you need a sugar rush, eat an apple or a satsuma – not pop, cakes, chocolate or even fruit juice. Or at least start to alternate.
- If you feel a bit empty, drink water! (When you retrain your taste buds, nothing is as refreshing as a glass of water. And for that matter, carrots and apples are really quite sweet.)
- Drink water with your meal. It'll help you feel full.
- Alternate coffee with a glass of water.
- Alternate coffee with decaf or herbal hot drinks.
- Eat meals on a smaller plate than you usually use. (It's polite/not wasteful to clear your plate, and even if you stop because you're full, then we all know that by leaving it a few minutes we'll be tempted to start grazing again as those extra chips/tatties/scoops of ice cream etc. are right there in front of us. Actively seeking seconds is much less likely.)
- Order from the children's menu. You won't go hungry.
- Try to introduce a vegetarian meal once or twice a week, or twice as often as you do now. Quorn comes in a variety of shapes and sizes, and is really rather tasty.
- Remember that fruit, once blended, still has some vitamin benefits but in many respects is no better for you than a Coke – so don't kid yourself it's healthy because it used to be an orange. Treat yourself (occasionally).

What we mustn't do, of course, is overreact if we have a blip and announce, 'Hey, I've ruined it and might as well eat all the cake!' Indeed, we shouldn't be on a diet at all, as diets are an infamous con. If you want to maintain a weight just under 14 stone (as befits, to pick an example entirely at random, a six-foot ex-rugby player), then you need to eat and drink like a 13 3/4 stone person – forever. Or you could eat and drink like a 15-stone person, but you'd have to run 10 miles a day every day.

The vast majority of people who diet successfully end up heavier than when they started within a few months – it's something to do with metabolisms reacting to the unusual scarcity of food! The famous supermodel Christy Turlington was once asked how she kept in good shape even in her fifties. The interviewer (rather hopefully, we think) asked her, 'Have you been hungry every day for all these years?' The model replied, 'Not really, I've just never eaten as much as I wanted to'.

Again, you don't need us to remind you that people who sell you fast food sell on taste, that processed food is bad for you and that people who sell packaged food fill it with sugar for taste and salt to preserve it. Again, efforts to halve the amount of crap we eat will at least move us in the right direction.

Organisational approaches

Workshops and information help, but who doesn't know this stuff by now? Better is to make the healthier food cheaper in the canteen, to provide water coolers around the site and free jugs of tap water on all canteen tables. Perhaps the one initiative that everyone knows about is the cheap porridge that was provided for construction workers building the Olympic village for the 2012 London Games. Noticing that many of the workers set off really early, and many stating they didn't have time for breakfast (or at least snatching something unhealthy, best they could, on the way), cheap healthy porridge was provided, and it did help with concentration, energy levels and accidents.

Other examples include providing free fruit and carrot sticks in abundance. Incidentally, people will pick up and eat twice as many satsumas if they are 'easy peelers' rather than those fiddly ones, so of course making sure to resource the easy-peel variety very much fits the main theme of the book.

Using similar psychology, an experiment carried out by Ogilvy & Mather tried to encourage fruit selection. Apples were made available at an entrance but so were cakes. The apples were easier to pick up, of course – being front and centre – but it at least allowed for choice. Choosing the healthy option is, of course, self-reinforcing and cues the person involved to self-label with the self-fulfilling 'healthy eater'.

What was clever, though, was that the employees were nudged to do this by the presence of a mirror behind the apples and cakes, as Ogilvy & Mather were aware of the prosocial impact of a mirror. Studies show that it's hard to steal from a charity box, for example, if you can see yourself in a mirror or if there is a picture of an authority figure 'staring' at you. (You may have seen cardboard cut-out cops around and about.) Apple selection increased significantly compared to a control group.

Social media

A growing number of books and research suggest that social media can be as damaging to the psyche as it is time consuming. Worryingly, many of these books are coming from 'insiders', who keep their own children's contact with social media to a bare minimum. Jaron Lanier's *Ten Arguments for Deleting Your Social Media Accounts Right Now* covers the arguments effectively. In short, the problem is that many of these media are harmful *and* are designed to be addictive. It's just an electronic version of giving out free drugs at the school gates. In short, do not send around a picture of your new trainers hoping for lots of likes and feeling crushed if you don't get many – just pop them on and go for a run!

Exercise

Exercise busts stress, builds resilience and body confidence and releases all sorts of helpful and positive brain chemicals naturally. The bad news is that there's no getting around it. We need to exercise, and we need around 2 1/2 hours of moderate exercise a week to 'sharpen the saw' to quote the 7th and final habit of Covey's *The 7 Habits of Highly Effective People*.

There are constant reports in the media about just how beneficial exercise is for long-term health as well as boosting short-term mood and energy. The good news is that evidence is accumulating that just short bursts of intense exercise are all that's required and that merely walking is equally as good for you so long as you do enough of it. This rather mirrors how we acted thousands of years ago with migrations and/or long treks to hunting grounds, followed by short bursts of chasing or running away! Certainly, if you average your 10,000 steps a day and augment that with just two or three 10-minute bursts over the weekend, you probably have it covered.

"The idea is a couple of quick calls at 3 mph to help boost energy ... you've been on here for two hours at full speed and are clearly just training for a marathon!"

Some excellent news: recent studies have shown that 'weekend warriors' who cram it all into one or two days are almost as well off as people who do half an hour four or five times a week. Researchers in the UK and Australia report that the weekend warriors are 40% less likely to die of cardiovascular disease and 18% less likely to die from cancer than the inactive. The figures for regular exercisers are better – but really not much (41% and 21%, respectively).

Stretching exercises and yoga that strengthen muscles and lubricate joints are also key as we get older. If these are designed to address specific issues in the workplace, then all the better. (The most decorated footballer in British history, Ryan Giggs, played Premier League football as an outfielder until he was 40 and claimed this was due to spending as much if not more time on this aspect of his fitness as running.)

Organisational approaches

Sponsored donations to charity for verified steps taken, bike-to-work schemes, gyms on-site and sponsored gym membership for people not based at head office are the obvious issues that everyone knows about.

Less well known are the findings as to why people take up bike-to-work schemes, which were uncovered by Britain's Healthiest Workplace research. They found that the biggest single correlation with cycling to work was the fact that there was a metric that people in the 'c-suite' took an interest in. This 'what makes my boss happy/what gets measured gets done' finding can of course be filed under 'it was ever thus'.

Again, the world of safety excellence has shown us clearly that saying you want it and actually wanting it are not necessarily the same thing at all and that a workforce can always tell the difference.

So again, senior management buy-in is utterly key. Genuine, followed-up buy-in, that is. Indeed, one of the findings of the Vitality study is that having a wellness budget is important in determining success, but that internal reporting to the board is around five times as important as that.

Another interesting finding takes us to nudge theory and the 'don't just educate; educate and facilitate' theme of the book. Second on the list was whether or not there was somewhere safe and secure to lock up the bike, and third whether there were appropriate shower facilities. Similarly, on oil rigs, gyms will typically have women-only sessions, as nearly all regular gyms do. A less typical innovation is to have over-40s or over-50s-only sessions, where a more gentle exercise regime can be undertaken without having to do it in the intimidating presence of the powerlifting club.

The importance of branding and publicising

Many organisations have schemes in place that aim to encourage the taking of exercise, but a recent study shows that in 80% of the organisations they surveyed, less than half the workforce knew that they exist.

For example, discounted gym membership is an excellent nudge, and a full 29% of the workforce who know about the scheme will take advantage of that. Sadly, with only 1 in 3 knowing that the scheme exists, that means that the actual take-up rate is below 10%. It is highly likely to be skewed towards the 'already health-conscious'.

Increasingly, and reflecting senior management commitment, organisations are beginning to properly brand these schemes as they would a new product line and use the same slick multimedia platforms to push them.

Exercise and the 'law of unintended consequences'

This is as good a place as any to consider the issue of unintended consequences, and many initiatives need thinking through carefully if they are not to backfire. Sponsoring a work's rugby league team may encourage people who wouldn't otherwise have played to take up the sport. However, rugby (especially league) is a brutal sport, and many will get injured! (Of course, many a muscle has been pulled by employees running about playing touch rugby for the work's team, or on a badminton court, but it's a lower attrition rate!)

Perhaps more serious is cycling to work. It's very good for anaerobic issues, but it's also the most dangerous way to travel to work per mile covered, and, in some locations, really quite hazardous indeed. (A micro-mort is how likely a person is to be killed per one million instances. The figure of 1 'micro-mort' per 5 miles of cycling doesn't really compare to driving, which has 1 micro-mort for every 300 miles. Public transport is similar. Then there are air pollution issues . . .)

Some organisations attempt to nudge people into getting the work-life balance right by switching off the server between say 7 a.m. and 7 p.m. Since many recent studies have shown that failure to switch off is one of the great health risk factors, this seems hugely sensible at first glance. However, field reports suggest that many people find the lack of flexibility difficult. Being able to take time off from 3 p.m. until 6 p.m. to do the school run and/or make tea, then work 8 p.m. until 11 p.m. is perfect for some people.

Similarly, banning the use of hands-free phones in cars is hugely sensible, as studies show the distraction involved makes a hands-free call about 75% as dangerous as the (illegal) hands-on one. But if the work schedules remain such that utilising time in the car between sites is the only realistic way of communicating, then something will have to give, one way or another.

Smoking

All companies have clear policies about alcohol. You don't drink at work and you don't come to work drunk. The days of four pints at lunchtime are long gone (well, mostly). Smoking is a different matter, and no organisation needs to educate anybody about the dangers. Each packet bought has a warning that says, 'This will kill you', or suchlike.

Organisational approaches

Some organisations have moved to a zero tolerance approach. Rolls-Royce, for example, have a policy that no one is allowed to smoke on the premises or indeed on company time at all. There's no nipping to the smoke shack out back for a quick one every hour or so. Basically, if employees must smoke, then it's pretty much taking a walk off the premises at lunchtime only. Patches and other support are provided to help, but there's no fudging.

Their director of safety, Dr David Roomes, explained that he got the board to agree that building a smoke shack is basically enabling employees to harm themselves. Not everyone is happy, of course, but David reports that, actually, pushback has been more limited than might be expected, explaining, 'Only around 20% smoke at all, and 90% of those are thinking of or actively trying to quit.' Again, however, a holistic, constructive approach is best as the following quote illustrates:

'I hate the job . . . I'd lose the plot completely without the occasional fag break and a quick bit of chat and catch up.' Speaking to this person illustrated the realities of ABC analysis perfectly. They knew that even smoking 5 a day carries a high health risk but they were more than happy to trade that for the daily sustenance of some social contact.

Drugs

Regarding illegal drugs, the hygge principle applies unless your company has a zero tolerance policy and a screening programme. A far bigger issue in recent years has been the massive increase in the widespread legal use of opiates. In some of the more deprived states in the US, up to a quarter of the workforce are working on legally prescribed medications, and accidental fatal overdoses of prescription medications in 2016 amounted to more than 15,000. Clamping down on such casual prescription has led to many, now dependent, individuals switching to illegal opiates. (See, for example, the experience of Huntingdon, West Virginia as documented in the Louis Theroux BBC film *Heroin Town* of 2017). Europe is catching up.

At the start of the book, we suggested that there are worrying trends that UK plc needs to address as part of its sustainability strategy (let alone a CSR strategy). This is one of them, and the more proactive work we can do, the better.

'Depression is when we think too much about the past in a negative way … we need to be more present … anxiety is when we spend too much time worrying about the future … we need to be more present.'

(Lao Tzu, famed 6th century Chinese philosopher)

Mental health

There are times when it's good to keep a stiff upper lip, but not to the extent that it costs you your health.

A young father called William (from London) joins his brother Harry in talking openly about when stress and/or a genetic predisposition and/or a trauma become a mental health issue. Led by many public figures, including the royal family, it's becoming less and less taboo, which is a good thing, as it's an increasing problem.

The extent of the problem

More than 6,000 people feel so wretched that they commit suicide every year in the UK, with men outnumbering women by 3 to 1. Around 5,000 are of working age. We have a particular problem with self-harm in the UK, we top the European charts, with 1 in 10 people self-harming through, for example, cutting themselves or taking tablets.

The chief medical officer of the UK estimated that the number of days lost to stress, depression and anxiety increased by 24% in the four years up until 2013. The UK loses 70 million sick days a year directly attributable to mental health issues. The HSE in 2015 reported that 'stress' accounted for 35% of all ill health cases that year. It's actually three times more common than cancer. Of course, many struggle in and 'present', and the Centre for Mental Health in 2010 estimated that 'presenteeism' (there in body but not in spirit) accounts for more lost productivity than absences.

This may well be the result of people being more comfortable reporting such issues, or may well reflect the increased workloads and precariousness of work following the financial crisis of 2008, coupled with technology changes that make it increasingly difficult to be 'off'.

Perhaps reflecting the impact of social media, the situation regarding young women seems to be particularly worrying. One in four young women will suffer with some issue related to their mental health each year (with the most common being anxiety and depression). Last year, 'stress' accounted for 43% of all working days lost due to ill health, and for 34% of all work-related ill health cases, yet 95% of employees cited a reason other than workplace stress for their absence due to stigma. Officially, both 'bad backs' and 'the flu' featured strongly.

What seems certain is that this is a serious issue for all organisations and that those that address it effectively and proactively will be better positioned than those that ignore the issue.

Shifting attitudes

More positively, the issue of mental health is becoming far less of a taboo topic and people are addressing it. The media is full of celebrities holding up four fingers in photos to denote that 1 in 4 figure. Poppy Jaman, programme director for City Mental Health Alliance, a network of City of London-based organisations that are raising awareness of mental health, says, 'The business case for addressing mental health and wellbeing has been established and is now featuring on many boardroom agendas,' adding that companies are increasingly seeking to support employees who are experiencing difficulties with mental health.

There is still a long way to go, though. As it stands, around 94% of business leaders admitted to prejudice against people with mental health issues in their organisation, and half of all employees state that they would not talk to their manager about a mental health issue. The MP Norman Lamb is striving to get the First Aid Regulations amended to take in mental health but hasn't succeeded yet. When you look at the figures, you have to ask, 'Why does he need to strive to get this through exactly?'

Another issue is that of in-house politics. Ideally, line management will own the issue, and the HR and SHE teams will work together to support them with that. It doesn't always work that way.

The Stevenson/Farmer review of mental health and employers quotes studies that suggest that for every pound spent on mental health employees recoup 10. They suggest that organisations address the following list, the more generic elements of which, we'd argue, are entirely congruent with maximising wellbeing generally.

1. Produce, implement and communicate a mental health at work plan.
2. Develop mental health awareness among employees.
3. Encourage open conversations about mental health and the support available when employees are struggling.
4. Provide your employees with good working conditions.
5. Promote effective people management.
6. Routinely monitor employee mental health and wellbeing.

Recently one of the authors ran a wellbeing workshop for a major international company and referenced the figure a few pages back, explaining that mental health first aiders' job is to spot people edging towards a critical level of depression and/or anxiety and refer them to the appropriate support. The job of a holistic wellbeing programme is to help proactively keep people in the middle! A senior manager in the audience put his hand up and said:

> hearing mental health discussed in this analytical way, as an unavoidable resource issue that needs a strategy and tactical solutions, like any other resource issue, is really heartening.

He went on to explain that no one in the room knew that his recent two week absence had been caused by anxiety and that he'd been anxious about the topic coming up in this talk. Hearing it described in this way had empowered him to share. The CEO responded by announcing she'd like to formally thank him for his 'bravery' and a spontaneous round of applause broke out. Things are changing for the better.

Help and initiatives

High-profile individuals in the media (and within organisations) can make a huge difference by talking openly about their personal issues, and it's very encouraging that this is something becoming increasingly prevalent. Whether it's a Stephen Fry, an Alistair Campbell, a Prince, the CFO of Virgin or just the Bloggs & Co. shift manager talking openly about their own issues, it helps empower everyone to step forward and open up.

At a recent SHE show event, a speaker talked about his own issues after being caught up in one of the explosions at the Europa Hotel in Belfast. Afterwards, a dozen people sought him out to empathise and share their own experiences, and as many others contacted him afterwards.

We'd estimate this represented about 10% of the audience and that this reflects the fact that people are feeling more empowered to talk about their feelings. Research indicates that talking to a friend or colleague who understands you can be as effective, or even more effective, than talking to a trained counsellor, especially if it is about something as specific as post-conflict stress in the armed forces (see TRIM, for example). One hugely welcome development is that the general principle is being adopted by civilian organisations, too.

Mental Health First Aid (MHFA)

Mental Health First Aid courses are an excellent way to raise awareness and develop champions within the business. Mental Health First Aid England encourage training as many members of the management team as possible in at least an awareness of the issue. They now have over 1,000 individuals trained to deliver this course (which costs just £69–200, depending on whether you want a half day, full day or two-day course), and hundreds of thousands have now been trained.

Courses highlight:

- the data above;
- giving some food for thought on how you might deal with issues; and
- contain exercises highlighting our instinctive discomfort and prejudice towards the topic.

Interestingly, but appropriately, the UK construction industry has been at the forefront of training first aiders, as, even in construction, workers are up to 12 times more likely to die from suicide than from falling from height.

Mates in Mind

Initiatives such as Mates in Mind explicitly seek to reduce stigma by making conversations about mental health part of the fabric of the organisation. Role models and champions are trained over a period of up to a week to lead this ongoing process, and every worker gets at least an hour's briefing as part of their induction. Volunteers and key individuals get half a day, and some can undertake the two-day course to become 'mental health first aiders'. Applied suicide prevention training may also be given.

Much of the work in this field in the UK has been based on pioneering work by the charity State of Mind, following the suicide of legendary rugby league 'Man of Steel' Terry Newton in 2010. Initiatives in rugby union and other sports have followed. The (anxiety-suffering) Kiwi/Leicester scrum half Jono Kitto articulated the benefits in the *Sunday Times* in March 2018:

> The fact that I could talk about it, not feeling that it was something shameful and weak, and that I could help others experiencing similar problems – it has helped so much.

Something of a nemesis for Welsh rugby over recent years, James Haskell has described 'struggling with confidence' and says of proactive work over the years to deal with this:

> I have a lot of tools that I can use if I'm having a bad day, not feeling particularly confident or getting hammered in the media . . . things in our head are 10 times worse than the reality.

He goes on to stress that the key is 'to sit down and talk'.

Of course, signposting to existing providers is essential if more than a 'good chat' is needed, and there are a large number of wonderful organisations offering to help. Both the Samaritans and MIND in the UK provide a 24-hour phone line and a national network of offices. Individuals may also find excellent help and support from Mental Health First Aid England, Remploy (who support people with disabilities in work), CALM (a charity seeking to reduce suicide in young men) PAPYRUS (seeking to reduce suicide in young people generally) and the Lighthouse Foundation (a construction-specific charity). This isn't an exhaustive list.

It's worth considering mental health charity MIND's five tips for mental health, available with a lot of other useful material on their website and reported here verbatim:

- Give. A simple act of kindness can go a long way to making you feel good too.
- Keep active. Keeping active is key to keeping our bodies healthy and making us feel good.
- Keep learning. Learn a new skill or take on a new challenge. Doing it will make you feel proud.
- Take notice. Take time to look at the world around you. It will make you feel differently.
- Connect. Connecting with others (family, friends, colleagues or neighbours) can give us a boost.

Similarly, the Mental Health Foundation lists 10 practical ways we can look after our mental health:

1. Talk about your feelings.
2. Keep active.
3. Eat well.
4. Drink sensibly.
5. Keep in touch (with friends and family).
6. Ask for help.
7. Take a break (even if only for five minutes).
8. Do something you're good at.
9. Accept who you are.
10. Care for others.

Research published by Oxford Economics (for Sainsbury's) in 2018 suggested that the factor that correlated most strongly with unhappiness was constantly eating alone. Other important factors included being time poor, tiredness and a poor sex life, but with the research confirming that social contact is key and face-to-face social contact especially important. (Talking on the telephone helps but not as much as talking in person.)

You'll of course notice a huge (and not coincidental) overlap here between both lists and with the strategies Martin Seligman and others propose under the banner of 'positive psychology'. Again, there is simply huge consensus about what we need to be doing as individuals and what organisations need to be facilitating and supporting. You'll also notice a clear overlap with the five factors discussed in the main introduction itself. This includes the one factor most people find surprising – indeed, the only one people find surprising – 'giving'.

The charity organisations listed above are mostly run by volunteers, but commercial organisations can provide an opportunity for all staff in-house through training up mental health first aiders and through seeking volunteers for a buddy system.

Buddy system

The buddy system, which is something in between Mates in Mind and a mentor, really helps too. A return to work after a period of absence due to MH issues can be daunting, of course, but having a 'mate' greet you at the door with a big smile can really help.

One interviewee put it like this:

> I was very worried that I'd be met with embarrassed silence or even teasing. So when my 'buddy' met me in reception with a broad smile, an arm around the shoulder and reassurance, it really helped.

The authors have worked with a lot of organisations over the years, and nearly all have said that their employees are their most 'treasured resource'. They haven't always backed that up with actions, and though individuals have always looked out for their mates and peers, it does seem as if a more systemic approach is being taken.

Several interviewees said that watching an employee stand in front of their peers and admit they had been struggling, and have that received sympathetically, has been as moving and empowering as it would have been unexpected a few years ago. To paraphrase the famous line from the film *Network*, it seems 'we're (all) mad as hell, but we're not prepared to take it any more'.

Followership

At this point, it's worth quickly introducing a concept that we feel should be included in any course where volunteering and intervening is important and that's the relatively new field of 'followership'.

In 1964, a New York woman, Kitty Genovese, was murdered outside a block of flats in which she lived. She was attacked twice. The first time, dozens of people heard the attack and opened windows to see what was happening, which drove the assailant away. None called the police, however, so when he returned to gawp at the crime scene, she lay injured where he'd left her still, and he resumed the assault. This horrible incident at least inspired some groundbreaking research into the bystander effect.

Experiments were set up to see how easy it is to 'say something first'. Subjects were asked to pick which of a group of lines matched another. It was difficult, but with only one mistake in a thousand. However, the study found that if you're the unknowing subject, last but one in a long line of stooges who lie and say that 'B' matches the control when it's actually 'C', then most people will deny the evidence of their own eyes and agree that 'B' is indeed the match. However, if just one person is allowed to break ranks and say that 'C' is the match, then the majority of subjects will say, 'I don't know what the rest of you are seeing, but I agree with X – I think it's 'C'.

It illustrates how reluctant people are to speak up (as on the night of the murder), but also, and this is key, how just one person can make a big difference to the behaviour of others around them.

If your organisation is going to address the issues in this book, you're going to need both volunteers and people who are happy to copy the good ideas of others! (The following short clip might amuse: www.ted.com/talks/derek_sivers_how_to_start_a_movement.)

Employee assistance programmes

Many organisations now provide an employee assistance resource. This might be face-to-face, or it might be by phone or Skype, and best practice is to open this up not just to employees but also to their families, as problems tend to be multifaceted and clustered, as above. Often, the help offered will be based on cognitive behavioural therapy (CBT), and frequently there will be a limit to the number of sessions an individual can undertake – usually 6–10.

Many are excellent, but some take good money off organisations to merely 'screen', which, in effect, means to simply refer people straight on to the health service and/or the organisations mentioned above. This can be done as effectively and more quickly in-house. One really bad example we found was that of a highly experienced and very well remunerated Harley St doctor, who provided an 'emergency response' service where he guaranteed to return a call within 24 hours. Then he simply referred the person on.

More typically, it's getting people to take advantage of the employee assistance resources that is of most concern. As Steve Hails, health and safety director for Tideway said when interviewed by the authors: 'In truth, use of the service is minimal, and we have to keep reminding people it's available.'

In the closing chapter, we suggest setting teams of end users to SWOT analyse key issues. We're suggesting the efficacy of any EA programme is an area to address.

The job itself   6

The previous chapter put the person as an individual at front and centre. It addressed what resilience is, how to build it, how companies can help with that, and it looked at the way organisations are responding to individuals in crisis.

Almost in passing, so far we've mentioned that 'good work is good for you' and that an individual's relationship with their direct boss is hugely important. (The reverse is also true by definition – bad work is bad for you.) We'd now like to turn our attention to what good work actually is before focusing on the non-technical skills leaders require.

We'd like to think that the material is very straightforward – which rather begs the question why so many organisations (and leaders) are making such a pig's ear of it and why Homer Simpson is the voice of his generation.

The elements of a rewarding job

As above, it's best to have a job in a field that you find interesting or you'll struggle to devote the energy and effort to be good at it, and it's difficult to find satisfaction in work that you don't think is adding value. We nearly all want to feel we're chipping in our fair share. Even footballers on hundreds of thousands of pounds a week check their 1–10 ratings in the papers.

For the rest of us, if we are engaged enough in our work for the day to pass quickly, then that's a really good thing. Technically, it's called 'flow' and reflects Einstein's user-friendly definition of relativity. He quipped that to a man kissing a pretty girl, a minute seems like seconds (there's a theme developing with his examples), but to a man with his hand in a fire, however, a minute will seem like hours.

People in a job where time always drags are likely to be developing stress, or worse. It doesn't help if it's a really bad job. This may involve working in the cold, the wet and the wind and long hours for little pay and/or a long commute. Or it may be working on a zero-hour contract for the likes of Uber.

These examples are self-evident, but the culture is often even more important. Imagine a forklift driver, Gill, who is struggling with all basic areas of her task:

- job design and environment;
- competence and knowledge;
- equipment; and
- rules and processes.

Gill has to move 50 pallets an hour or she'll be in trouble. Technically, this is just about achievable – indeed, it's based on time and motion research. Unfortunately, the sample run was in a perfect environment at the HQ testing bay. The yard in which she actually works is crowded with pedestrians, other forklift trucks and lorries. The movement of these visiting lorries isn't controlled in any meaningful way, and it's dangerous and chaotic in the yard.

Her training was perfunctory, so effectively she had to learn on the job, which was hazardous, but at least she got away with nothing worse than a few near misses. Her forklift truck is old, temperamental, battered and badly in need of repair and service. The old hands have long ago learned to carry their own toolkit, but she's not much of a mechanic, and she has to ask others for favours, as the engineering team are firefighting constantly. The very idea of planned maintenance is a joke. She can, of course, park up and wait, but her basic wage is terrible, and hitting those 'stretch' targets is the only way to make a living wage.

The trouble is that asking for favours gives the old hands plenty of scope for ribald and misogynistic humour. The culture is one of bullying and blame, and complaining about the worst of the comments has got her labelled a humourless troublemaker. This is important, as many writers (see Wilde, HSW, May 18) suggest that the most important factor in workplace wellbeing is the psychosocial environment.

Rules are mentioned only when something has gone wrong because 'blind eye syndrome' is rampant, and the golden rule is the basic: 'just get it done'. Gill would ask about training, career development and about job rotation, but she knows there's no point, and anyway, there's no one to ask. There is one manager, but the only interaction she's had with him (or anyone else) since arrival was when he slowed his car, lowered his window, smiled and gave her an inquisitive thumbs up. She thought after that he'd have stopped if she'd asked him to, but she of course smiled back and gave him a 'thumbs up' back.

We can all agree that this is not a 'good' work situation. The day does pass quickly but only because she has to move flat out at all times, and if she has any predisposition to mental health issues they will flare up sooner rather than later.

Basic frameworks

There are lots of taxonomies that are rather more refined than the basic four above. HSG 218 (Managing the Causes of Work-Related Stress) suggests:

- Demands. This includes issues such as workload, work patterns and the work environment.
- Control. How much say the person has in the way they do their work.
- Support. This includes the encouragement, sponsorship and resources provided by the organisation, line management and colleagues.
- Relationships. This includes promoting positive working to avoid conflict and dealing with unacceptable behaviour such as bullying.
- Role. Whether people understand their role within the organisation and whether the organisation ensures they do not have conflicting roles.
- Change. How organisational change (large or small) is managed and communicated in the organisation.

It's an excellent piece of work, free to access (see http://www.hse.gov.uk/stress/ for the latest link). There are two ways to use it. One is to monitor reactively to ensure that we don't let things get so bad that we break employees entirely (because we may even get prosecuted). The other, suggested in this book, is to use it as the basis of proactively redesigning tasks to maximise the amount of flourishing and thriving that's occurring.

With the latter in mind, Peter Warr's vitamin model describes a framework for what the perfect job might look like for you individually. Although we like to keep to triptychs whenever possible, you'll see that there are some elements, not mentioned above, that may be important. The model is a 'vitamin' model because it suggests that too much of a good thing can tip over and become a bad thing.

The factors are:

1. Money. As above.
2. Opportunity for interpersonal contact. Also as above, but from an organisational perspective also when discussing leadership non-technical skills.
3. & 4. Opportunity for skill use and variety. Self-evident elements, with obvious links to meaning, achievement, coaching, flow, delegation, etc. Though we say 'self-evident', a recent example from the UK shows how a lack of these elements, even on a specific day, can have severe consequences. After the Manchester bomb attack of May 2017, fire services personnel were, as it transpired, unnecessarily held back from attending to victims for more than two hours, as the scene had been declared 'hot' (meaning too hazardous, as the attack might be ongoing). In May 2018 the fire service reported a surge in stress-related absence and take-up of counselling services in the previous year as their firefighters struggled with feelings of 'guilt' and frustration.
5. Valued social position. It's a primeval drive to be considered useful, so we're not likely to be pushed out of the cave any time soon. That basic drive aside, it's a very personal thing, and we might – very briefly – consider Brexit and immigration, as it illustrates the individual perspective. Some people may look at a menial job and think, 'I'm not doing that', even if it's all they are qualified for. Others, coming from a different mindset, will instead think, 'A job! A real, paying job. Excellent!' and they'll stride from their front door with pride, even at 6 a.m. Naturally, employers respond to these attitudes and the behaviours that flow from them.
6. Opportunity for control. Some people are control freaks and hate being told what to do by anyone or anything ever. They even swear at the sat nav. Others have a greater tolerance and prefer a little direction – indeed, they may find too much autonomy positively stressful. For example, people who have chosen to work in uniforms can be quite low in this need, and people who choose to be self-employed can be quite high. A practical day-to-day example of best practice: a toolbox talk about the day's tasks should be 25% dialogue at least, allowing those who would like to say something a genuine opportunity to do so.
7. Goal and task demands. When we think of the FLT driver described above, it is self-evident. The next section covers coaching and motivational interviewing, which stresses that the very best goals are hard but realistic and set with the person him or herself. A key model here is that of optimal pressure. If goals and task demands are too high then of course we're more likely to be stressed and can suffer 'burnout', but if they are not high enough

then we are likely to be a little listless, demotivated and performing far below our potential (or 'rusting up').

Task identity and traction is important here as well as links to 'meaning' generally. It reflects how well we understand how what we're doing fits in with the general picture. A famous example would be at NASA, when in response to the question 'What are you doing?' someone sweeping up told the visiting John F. Kennedy 'helping put a man on the moon'. Again, toolbox talks and frequent communication with supervision is key here.

8. Level of uncertainty. We don't want to be told 'can't say' when asking about rumours, we don't want to be wondering if our flakey and moody manager will be in a good or bad mood today, and we don't want to be that forklift truck driver working in a blame culture, as above, who knows that one thing they really need to avoid the sack is an extended run of good luck.

9. Physical security. In some respects, this is an obvious one, reflecting the Maslow hierarchy's most basic need. By this, Warr doesn't just mean not being involved in an accident or developing a bad back but also not being in conditions that are physically stressful, such as heat and noise. The world of safety has long known that in the middle of a busy shift, a worker will do many things to make life more comfortable, such as not wearing a hood that's heavy, cumbersome and fogs up so you can't see what you're doing. It's annoying and slows you down. Failing to wear it might cost you your eyesight, your health or even your life but probably not today.

Exposure

This is a good time to discuss the biggest issue facing us in the UK, which is such a difficult problem that few even acknowledge it. It's an issue that makes many corporate social responsibility proclamations something of a bad joke.

The British Occupational Hygiene Society (BOHS) stresses that for every 100 deaths in the UK caused by work, 99 will be because of long-term health issues caused by work. The Rushton Report suggests a deliberately conservative figure of 13,000 deaths annually, which is not decreasing as the control of asbestos exposure works through but is increasing because of nanotechnologies.

It costs the UK around £14 billion a year. Of this, around 20% is born by society, including the NHS, and 25% is born by organisations. However, politicians think in five-year cycles, and the boards of companies, with stock-holders in mind, think in even smaller time frames. It's ABC analysis raising its head again. A full 55% of this cost is born by the families.

Steve Perkins, former head of the BOHS, implores organisations to take notice of these figures and to get involved in the Breathe Freely programme. Some organisations are indeed taking heed, and many or the bigger, more clued-up organisations are now insisting that the supply chain organisations provide screening. For example, Tideaway are piloting a scheme with the BOHS for a certificate in controlling health risks. Ongoing screening can, of course, deliver variable data regarding noise, dust and vibration, etc., but it does at least raise awareness and encourage the reporting of issues.

What all organisations can do, however, is to roll out toolbox talks that cover this data and to stress that if anyone raises an exposure issue, despite a perceived lack of urgency, it's highly likely to be highly important and can't be left to one side until 'a suitable time' because there's a good chance that'll never happen, as only the important and urgent issues ever get looked at. Better still, sign up to Breathe Freely and set up an empowered workforce project team to proactively seek out potential issues.

Action teams

As well as setting up research teams to look at exposure, similar teams can be set to work assessing any or all of the above lists. Here are some suggestions regarding issues that experience shows are often rather suboptimal and are grouped under the HSE guideline headings.

Conflicting demands

Is it, given the time and resources available, possible to work both effectively and safety? If it isn't, where are the pinch points, and what needs to be done to overcome them with something more constructive than 'pay your money and make your choice'? More specifically: Is presenteeism prevalent? Is 'leaveism' prevalent? (This is increasingly likely since the financial crash, where people work in their own time such as evenings, weekends or even holidays just to keep on top of things).

Control and change

How much say does the person have in the way they do their work? Are changes clumsily imposed on employees so that the false economy of more expensive

"It says here that people often self-select into
jobs that best suit their character ...
would you two gentlemen care to comment on
that for this dissertation I'm writing?"

retrofit adjustment is standard? Do toolbox talks and weekly briefs contain a large element of dialogue so that, as above, the actual experts get a chance to share that expertise?

Support

Are attempts to use new skills and/or technology followed up with monitoring and coaching or are employees left to get on with it? Are there mental health first aiders proactively looking for warning signs that people are struggling?

Relationships

Is bullying part of the culture? Is blame part of the culture? Do the two coexist when someone is blamed for complaining about bullying? Is development, empowerment and coaching part of the culture? Are people frequently caught 'getting something right' or is praise a rare event?

Role

Do people understand where they fit into the big picture? Do they understand not only what they need to do but also why they need to do it so that they can show 'operational dexterity' when something changes?

A simple model of organisational culture says that once diminishing returns have been reached regarding systems and procedures, then moving forward is dependent on two factors that cut across all of these subdivisions. The first is a mindful learning culture that accepts that there will always be problems and issues to deal with and that weaker cultures reactively wait for the problems to find them. Stronger cultures proactively seek out the problems.

The second factor is frontline leadership. The research shows that transformational, rather than transactional, leadership delivers the best results. This means:

- praising rather than criticizing;
- empowering whenever possible;
- coaching rather than telling;
- communicating effectively with passion (which means two-way communicating, not just using simple and powerful language); and
- leading by the right example (all leaders are leading by example at all times, whether they want to be or not).

Work groups looking at these issues will always come up with ideas, suggestions and enhancements that will make 'doing a good job quickly and safely' more viable. The changes that result will help wellbeing directly, and being involved in the process is itself hugely beneficial. It's been suggested that, if asked well, the only question a manager needs to ask is 'What do you need from me to be able to work quickly without hurting yourself?' The Yorkshire ex-head of IOSH, Gerard Hand, quips that he simply asks, when addressing safety specifically, 'What's dodgy about this job?' I'm sure he'd agree the equivalent of the wellbeing question is 'What does your head in about working here?'

It plugs directly into a win-win long-term investment rationale, so a warning: if an organisation is hoping for only magic bullet quick fixes, then they perhaps shouldn't set up such teams, as it'll open up a huge can of worms and create hopes and expectations that will be dashed.

If, on the other hand, you have the management commitment to action anything that appears 'high impact' (there will be lots of that), then it's the very definition of 'investing in . . .'. This applies even if you only action 'high-impact, low- to medium-cost' suggestions. It'll still be a big step in the right direction. Of course, organisations should properly cost-benefit analyse the 'high-impact, high-cost' suggestions, share the rationale behind the resulting decision and action a few 'medium- to low-benefit but interesting and low-cost' suggestions for symbolic and political purposes!

The next section, to use an annoying modern expression, drills down into some of these key transformational leadership behaviours in more detail.

> The first question shouldn't be the reactive 'Is this (now ailing?) person fit for work?' – It's to proactively ask: 'Is this workplace fit for a person?'

Non-technical skills

7

Above we've covered approaches to deliver clear thinking and effective decision making. We turn now to non-technical skills to enhance communication and team work and also to generate a positive and supportive environment. Readers may recognise the combination of the two as the essence of 'crew resource management' (CRM) most frequently used in safety critical environments such as aviation. The motivational role of empowerment and the interplay between mindset and clear thinking are key messages of this book.

Some aviation companies (EG CHC Helicopters) have rolled out CRM to *all* staff, and we couldn't agree more because the essence of the book is that it doesn't have to be news- headline worthy to be utterly critical.

The transformational leader

Working for a transformational leader is excellent for an individual's wellbeing! This section looks at some techniques and practical issues around:

- empowerment;
- coaching; and
- assertion.

Studies show that transformational leadership correlates with safety excellence, with transformational leaders tending to:

- praise rather than criticise;
- empower whenever viable;
- coach rather than tell;
- treat people like adults;
- give and receive feedback calmly and objectively; and
- know they are leading by example at all times, so strive to do so positively.

These traits also underpin a work culture supportive of wellbeing.

The first thing to say about soft skills or non-technical skills is that, as ever, it's not entirely about the skills themselves. To a great extent, it's actually about subconscious unspoken norms and perceptions that surround them.

There have been many recent articles about Brexit and Donald Trump that have sought to understand what motivated tens of millions of people to vote as they did, even though objectively it seems against their best interests. Leaving aside all the politics and economics, several social scientists have focused on two core societal building blocks that massively influence our perceptions and, therefore, our actions. These are reciprocity and liking.

We are, by default, and historically speaking, for very good reasons, genetically wary of 'other'. The 'them' we don't know. This might be refugees of a different colour, but it could also be 'them in suits' from head office; it's certainly 'them from the next town' and could well be just 'them on the night shift'. It's a universal emotion. For example, one of the authors presented at an event in New Zealand in 2016 and was advised to make sure to use the term 'Jaffa' for comedy effect – it meaning 'just another fellow from Auckland'!

Trust is the key issue that underpins the psychological contract, but few of us are hardwired to trust people. It has to be earned, and building it takes an awful lot more effort than losing it because we all know that it can be lost in a moment.

We find that the only way to overcome such instincts is to meet these people and get to know them. For example, the 2016 Palme d'Or-winning film *I, Daniel Blake* by Ken Loach is, in essence, a film about an old, unemployed man living on disability benefits who befriends a young, unemployed single mother working as a prostitute who is also living on benefits and well on the way to becoming what's known in the tabloids as a '4 by 4' (4 children by 4 different fathers). As an outline, that doesn't sound too promising, we know, but if you watch the film, however, it's impossible not to fall for both characters and empathise with them almost entirely. The civil rights icon Rosa Parks wasn't the first person arrested for refusing to give up her seat in Montgomery but was the first well-known person from her work on church committees and as a seamstress. People knew her and they mobilised. In the UK the provocative *Daily Mail* headline

Instantly Bob's non-technical training course in the importance of avoiding mixed messages made perfect sense to him ...

'Murderers' was hugely instrumental in leading to justice in the case of Stephen Lawrence and in the setting up of the MacPherson Inquiry, which is held to have helped change UK culture generally for the better. It was the former editor of the *Daily Mail*, Paul Dacre (often dubbed 'the nastiest man in Britain') who wrote, and insisted on, the headline himself. An unlikely hero in such a story perhaps, but Stephen's father Neville once did some plaster work at the editor's house. Dacre met him and liked him.

We can get over-comfortable with friends and family at times, but acquaintances tend to bring out our best selves.

So a key element of the power of non-technical skills is that, at heart, it's about encouraging habits that lead to empathy and understanding. Genuinely listening to people, treating them with respect and seeking to maximise their potential breaks down the 'them and us' and maximises the chance of engagement and discretionary effort.

The second element, and perhaps the most powerful behavioural driver of them all, is fairness and reciprocity. Study after study has shown that being unfair is much more harshly rated than breaking laws and rules. After all, most laws are just attempts at catch-all rules that ensure fairness.

So, anyone making a half decent effort to listen, to empathise, to respect and to seek to help you maximise your potential simply obliges you to respond in kind the best you can. Well, usually. Behaviour breeds behaviour, not just because 'Do as I say, not as I do' doesn't even work on children but because it's fair. It's what's called an unspoken 'psychological contract'.

This isn't new, of course, and Dale Carnegie's classic book *How to Win Friends and Influence People* has been back in the bestseller lists because there are tips and advice in the book we still fail to apply as often as we should.

Almost every organisation the authors have ever surveyed has come back with 'management non-technical skills are weak'. Wherever an organisation is at present, though, the 'grift' principle applies. Double the number of times the following non-technical tools and techniques are applied and that organisation's culture will be transformed.

Ownership and empowerment

The Campbell Institute White Paper assessed the benefits of out-and-out bribes to get people to get involved in wellbeing programmes. Results were mixed at best, of course, and, again, the lessons from safety-based incentives are useful here.

The 'induced compliance' paradigm of dissonance is well understood and largely explains the highly contentious issue of incentives for safety. In a classic experiment, it was shown how a small bribe to vote for a disliked political candidate causes a shift in attitude in favour of that candidate – whereas a large bribe led to no such change in attitude. This is because, in the latter case, the behaviour can be entirely explained by a large external factor with no dissonance (I'm a penniless US student in the 1960s and, frankly, I need that $30). However, in the case of the small bribe, there was an element of needing to internalise the belief they were induced to express so as to make the subject feel less easily 'bought'!

In safety, it's consistently found that the very best inducements are small and symbolic and supported by rational coaching techniques, using questions, data, illustrations and, most importantly of all, utilising involvement and ownership in the process – all of which are explicitly aimed at getting individuals genuinely to internalise the value or message. The same is true of wellbeing. Gimmicky initiatives will convince few. Training modules that are delivered (cheaply) by computer simply aren't going to engage an individual like a more personal approach.

It's also key that we are genuine. The words 'integrity' and 'authenticity' are mentioned more and more often at conferences and in literature and, because of social media, getting it wrong will be shared widely and quickly by Generation Y, which is more entitled and less deferential than ever before. A positive side of this new reality from a learning perspective is that anonymous forums and 'ideas boxes' allow others to comment, build on arguments and even vote on the best ideas. It's excellent data and feedback if you seek to learn from it.

In short, whatever we do under the wellbeing banner, we need to be doing it with people, not to people.

Leading by example

To develop the authenticity theme, we need to discuss leading by example. In the world of safety culture, we've long known that any example of a leader failing to follow their own advice is disastrous. With peer pressure, there's a tipping point of around 90% where the new starts and subcontractors will fall under a lot of (often unspoken) pressure to stay with the majority. With leadership, 90% is a calamity.

This is a huge problem where management is unaware of the research showing that working more than 40 hours is pointless and are visibly working themselves into an early grave. If the general culture is brutal, then the best individually based resilience work in the world is just damage limitation. Worse, if there is an active wellbeing approach and an implication that people not actively on board can't, by definition, care enough about their wellbeing, then this contradiction is, at best, exasperating. (The famous psychologist R.D. Laing said that the best way to drive someone insane is to tell them that you love them and then behave as if you don't. You can say one thing and model another, but it doesn't work.)

A positive example, the current (2018) CEO of Anglia Water personally lost 1 1/2 stone in weight to show his commitment to their health initiative. He commented that it really added credibility to the questions and challenges he raised in the canteen about the availability of healthy options and why some things hadn't been changed enough. In America, the current (2018) United Rentals CEO pledged to lose 25 pounds or donate $25,000 to charity.

The CEO also has an important role to play in embedding approaches within the line. Managers may well complain that they have enough conflicting demands to juggle without 'wellbeing' being added 'on top'. It's vital that: (a) it isn't merely plopped on top and (b) the c-suite makes it clear that this is just a key element of the eternal push for excellence and sustainability – and a cost-effective one at that. Then: (c) ensure it is resourced and coordinated properly.

Coaching

A famous Stanford University study by Professor Carol Dweck shows that if you praise the effort rather than the innate talent of the person, the results are extraordinary. People praised for being clever will enjoy that, of course, but show the tendency to avoid challenging tasks in the future because trying and failing might undermine their self-esteem. On the other hand, people praised for their effort, determination and resolve are far more likely to embrace these challenges, as, succeed or fail, their positive self-image isn't at threat.

Reverse psychology (i.e. 'you're useless' so 'go on prove me wrong') may work for some, but praising hard work and mindset works for all. This is the grift principle again. Warm but vague words about 'our wonderful, talented staff' are better than the reverse but get no one anywhere.

A sporting example: A football coach might seek to use a bit of 'positive labelling' to his approach. He might say, 'Don't worry, Diego, you'll start scoring again soon – you're a natural finisher,' but it's better to say,

> I really admire the way you keep busting a gut making those far post runs match after match, even though you've had no luck at all for weeks.

It's obvious you know from your extensive experience that any match now you'll get a handful of chances and a hat-trick.

He might add, 'I know you're really mindful that the young players coming through look up to you and that you always try to model the right attitude.' This might just work, even though it's only true when cameras and/or scouts from bigger clubs are in town.

He may well walk off thinking, 'I wish we could sell this useless old goat and buy someone half decent', but he's done his job as a coach at least!

The best coaching is where a person we trust (ideally a mentor who models all the things we aspire to) leads the person to set him or herself a challenging but achievable goal articulated as a SMART goal (SMART being specific, measurable, agreed, realistic and time-set). Then, they monitor, support and coach as required.

'Nudge theory' or 'behavioural economics'

We've mentioned this briefly above, and will consider it in more detail here. Nudge theory (see the book of that name by Sunstein and Thaler) can be defined as making a change to the environment that is clever, cheap, based on an understanding of psychology and physiology and is validated. Positive labelling is a nudge.

The most famous nudge is the painted fly on the toilet in Amsterdam Airport, which men can't help but point at, with a reduction in splashing of up to 80%, leading to an associated improvement in cleaning costs and the environmental impact of cleaning materials. Crucially, this idea came from the man tasked with keeping the toilets clean. This is not coincidental.

It has been suggested that nudge is just manipulation – clever ideas for getting you to do what the organisation wants without you realising. Well, yes, but then flattery achieves much the same objective, and no one is going to ban that. It's also suggested that nudge theory has been around forever, and indeed it has. The famous 'Hawthorne effect' study was originally about which colour we need to paint factory walls to get the employees to work harder. It found that if you make a change it gives a boost and that it didn't much matter whether the wall was painted red, green or yellow. The crucial factor was that it was given a new colour. That said, the colour of walls can be important. The bestseller *Drunk Tank Pink*, by Adam Alter, for example, shows that using different colours can increase and decrease aggression levels. Similarly, lovely green forest scenes

AS HIS COLLEAGUES COULD TESTIFY, BOB'S COACHING STYLE LEFT A LOT TO BE DESIRED

can be calming, and many a sports club's visiting changing rooms are a very different colour to the home side.

Here's a lesson directly from the world of safety and leader/member interactions. In this case, organisations realised that systemically ensuring we stop asking 'Why did you switch off?' and substituting the question with, 'Is everything that caused you to switch off now under control?' led to a better outcome. This is because the former says, 'You had better have a good reason for that', which can directly lead to learned helplessness, whereas the latter communicates an empowering, 'Of course you did, or you wouldn't have done it . . . what happened?' That's highly impactful but free, so a nudge.

We discussed the meaning of 'safely, but by Friday' above and applied it to wellbeing; it's simply a case of ensuring we communicate, 'Of course we want you to work hard and make us lots of money, but not at the expense of your health.'

People believe and, infamously, obey people in crisp white lab coats (see the Milgram conformity studies), so how important information is disseminated is at least as key as the quality of the information itself. A Plymouth University study showed that the effectiveness of a briefing is as much about where and how it is delivered as it is about content. It's about a smartly dressed person, in a professional environment, delivering a crisp talk backed up with good-quality materials.

Consider handwashing. It's a really simple behaviour that costs billions of dollars and millions of lives annually, especially in healthcare, which is an industry notoriously poor at learning (see *Black Box Thinking* by Matthew Syed). Advertising agency Ogilvy & Mather was tasked with getting South American meat workers to wash their hands thoroughly because contamination rates were causing millions of pounds worth of products to be rejected upstream. The clever solution was to stamp hands on entry to the factory with a dye that only really thorough washing could get off. Huge savings ensued. They charged a fortune for this, we imagine, but it could have come from the frontline for free.

We'd like to suggest that, following some basic training in 'temptation' analysis and other basic human factor issues, any organisation can set up its own 'nudge' unit. They have a head start on the outside consultant as they already know exactly what's happening and why!

Picking up on the above, both the way messages are communicated and the way questions are used are excellent projects for understanding how the organisational culture operates and the simple changes that can be made to lead empowerment and wellbeing.

Feedback

It's important not to confuse delegation with abdication. Empowerment is largely based on choice and autonomy, but that's not just allowing people to 'get on with it'. Guidance, coaching or even one-to-one mentoring will also be required as part of a transformational leadership approach. Giving good feedback and active coaching is central to doing this well.

Imagine trying to learn to play golf but blindfolded and with no feedback as to where your ball has gone. You'd never improve. How about that exercise bike, treadmill or rowing machine in the shed you used to use but stopped once the counter packed up? Just like that exercise regime, without feedback we tend to grind to a halt.

Basic feedback

When someone is observed doing something constructive, then any praise that is soon, certain and positive, when delivered from a credible source, will be rewarding and will help reinforce the behaviour. The more of this, the better, and books such as Ken Blanchard's *The One Minute Manager* can help. Perhaps the key advice in the book is to try to 'catch a person doing something right'.

Whether it's volunteer behaviour, like putting a name down to be a mental health first aider, or where a behaviour has been recently enabled (perhaps making phone calls from a treadmill), we'll still want to reinforce it to ensure it embeds and becomes 'what we do around here'.

There's a famous old exercise in which a volunteer is blindfolded and tasked to throw a handful of marker pens or spoons into a bucket from 10 feet away. In the first run, they get no feedback whatsoever. In the second, they receive only negative feedback, and in the third all the feedback is positive. They typically do badly in the first run, worse in the second and may even refuse to continue if jeered with enough gusto. But they perform better in the third.

Praise is 20 times more effective than criticism in changing behaviour, and catching someone doing something right is a 'soon, certain and positive' payoff. On the third run, a thrower will instantly improve and will typically hit the bucket in 6–10 throws. However, this is still of course entirely person-focused, and we've added a fourth iteration here that demonstrates this.

Told that the rules remain in place and that the person throws the pens one at a time, blindfolded and from 10 feet, but that everything else is open to a design or teamwork solution, we soon have funnels and catchers and coaches suggesting a lobbed throw to help the catcher.

This is 100% successful because while praise may be 20 times better than criticism it's no substitute for designed facilitation. Encouraging people to walk around the ornamental lake is great, but if you're based in rainy Manchester you're going to have to buy treadmills!

Giving negative feedback

By far the biggest issue with negative feedback is either that it's the only sort used or, in many organisations, the utter lack of it – known as 'blind eye' syndrome. This is an issue, of course, because what we fail to challenge we effectively condone. Challenging is uncomfortable for many, so we avoid it if we can as ABC or temptation analysis rears its head again. 'It's OK to be challenged and it's OK to challenge' is an expression often heard in an organisation desiring a strong culture, and it must be a key aim of any organisation seeking excellence. Training people in these skills is the easy part. Creating an environment in which they are used readily is the hard bit.

Specifically, when giving negative feedback, it's vital to follow some key rules because if you break them the impact will be the opposite of what you intended. They won't be thinking of the behaviour at hand and any associated risk; they'll be thinking of you and not in a nice way. If they can, they'll let you know on the spot, usually through their voice tone and body language but sometimes more directly. If not, they'll just reduce the amount of discretionary effort, which is exactly what we don't want when we're seeking to empower. Often it'll be both.

The golden rules of giving negative feedback are:

- never personalise;
- never generalise; and
- never berate someone in front of others for extra impact.

A loud 'You're always doing this, you idiot' is a funny line to put in a 'Don't do it this way' training video, but only there. We rarely do anything 'always', and if you call me that, the only thing I'll be thinking about is its inappropriateness. If you do it in front of my peers, they may laugh, especially if it's frequently true. But I'll spend the rest of the day fantasising about having you killed. Worse than that, if I'm popular, the colleagues won't laugh. They'll be mortified on my behalf. If you look up 'workplace bullying' in any HR policy, you'll find that being criticised in front of colleagues will be high on the list.

Here's an example that has stuck with one of the authors over the years. The chap involved was an excellent leader: fair, resolute, consistent, thoughtful, clear in his communications – everything a leader should be. A former rugby league player, he was a key figure in a project that saw accident rates cut to a tenth of previous levels nationwide in 18 months.

One of the things he knew how to do was to give feedback for maximum impact and minimum unintended consequence. The incident the author remembers was at a residential training event, where one of the employer's trainees was rude to a waitress. It was nothing too serious but was definitely out of order, nonetheless. The author's chap saw this but said nothing, waited a minute or two, then casually asked the young chap if he could have a quick word about something that had just occurred to him. Unconcerned, the young man followed him out of the room, but when he returned five minutes later he looked shaken and white. Before he left the room, he gave a sincere apology to the waitress.

Normally, tactfully taking someone to the side and sticking objectively to the facts at hand is all that's required. That's exactly what this man did. He didn't personalise, generalise or raise his voice. But this clearly didn't stop him from articulating his observation with impact.

In doing so, he'll have done that young man a service. And that's the point. Negative feedback is sometimes required, with even gurus of positive psychology such as Martin Seligman suggesting a 3:1 ratio. Whatever the ratio, doing it well is essential. Many an individual avoids the difficulties here by rarely doing it all, and then, like a backfiring motorbike, charging in unadvisedly all guns blazing.

BIFF, BIFF (a second time), then BIFFO and maybe even BIFFOF

These simple acronyms help anyone remember how to give negative feedback, especially if we're angry and struggling to operate from that rational prefrontal part of the brain. It stands for:

- Behaviour – this is what you did, causing . . .
- Impact – and this is why it's suboptimal
- In Future, we would like this behaviour instead
- For this (different) output.

It's BIFFOF because, after repeating it once, if appropriate, but certainly no more than once we really need to upgrade to:

• Or this will happen.

This needs to be Followed, though, of course (F), or all credibility will be lost. I suspect from the lack of colour in the apprentice's cheeks that the chap above went straight for BIFFO and was all the more impactful for being delivered calmly.

Feed forward technique (and creative problem-solving in teams)

This is useful in team situations where you don't have the seniority to insist people take your feedback. It's a simple technique, and any team that uses it well won't be going far wrong. The idea is that improvement ideas are sought and articulated in a positive way, so to pick up on the above example where someone got it wrong, you may ask if it's OK to make a suggestion, then, if it is, say something such as,

> He got the message OK, but I'd like to suggest you'd have greater impact in the long term if you took him to one side to give him that feedback. I worry that he's not thinking about the feedback; he's just thinking about you and not in a good way.

The trick is that the person getting the feedback is obliged to do one thing and one thing only, which is to treat it like a present. So we first say, 'Thank you' and then either: (a) pop it in a drawer; or (b) use it. As you can well imagine, it plugs directly into, or at least overlaps with, any sort of group problem-solving. Ensuring that everyone gets to take a turn and is listened to patiently ensures that the loudest and/or most senior person in the session doesn't dominate. (The new analogy is to not be a HIPPO, which started out as 'big mouth, small ears', but we now believe it's been embellished to 'Highest Paid Person's Opinion'.)

Active listening

You'd think that the skill of listening wouldn't take long to master, and it's true that the component parts of an active listening technique are not difficult to grasp. It's using them in action that's tiring and difficult. The component parts are as follows.

1. Pay full attention

This is not just turning off your phone, not looking over their shoulder and not cutting them off rudely. Nor is it simply waiting for them to finish speaking so that you can say your thing. It also requires actively listening to what they are saying by also watching their body language and remembering that 85% of communication is in the tone and body language.

2. Paraphrase back what they have said

You confirm you listened by repeating it back using slightly different language, showing you have processed it rather than merely parroting it. It's also important to paraphrase back the meaning you got from it, if there's anything in the 85% mentioned above that jars or needs clarifying.

This is key. If they are saying something but not sounding as if they genuinely mean it, then this is the time to clarify. This doesn't have to be confrontational. Something like, 'John, I'm getting the impression that you have concerns about X & Y . . . could you clarify that?' suffices.

3. Cut the boxing ring down with a 'yes/no' in the face of evasion

If you feel you're being fed a mixed message deliberately, perhaps to pass the problem onto you, then 'yes/no' is your friend. As above, saying 'safely, but by Friday' suggests 'by Friday', so a way of clarifying that would be to challenge by asserting, 'I can do it safely, and I can do it by Friday, but I may well not be able to do both. What I'm hearing is that, given the choice, you want this out by Friday, come what may?' It's likely that our supervisor will bluster, 'Er . . . I didn't say that.' Then, to drive home the point, you will need to ask, 'So, to clarify, if I need to delay until the weekend to ensure no corners are cut, is that OK?'

You may well get an, 'I didn't say that either!', but they are in a situation now where they have to either say one or the other, or clearly refuse to say either. In any inquiry following something going wrong, the initial 'get out' that they were trying to set up, consciously or otherwise ('I explicitly said do it safely . . . I'm not sure what the problem is'), is now removed. It will be clear to all now that they didn't say anything explicit and tried to fudge it and pass the problem on to you.

Assertion

Assertion is not asserting yourself; it's insisting that your rights be respected without impacting on those of others. Being able to give negative feedback well and to challenge mixed messages are key skills of assertion. An assertive 'argument' might go something like this:

> With respect, boss, I think X decision is wrong. What about Y factor?
>
> That's a good point, and actually considering factor Y did indeed give me pause for thought. However, on balance, I still think we should do X, and I get the casting vote.
>
> OK, I'll try not to say 'I told you so' if you're wrong. But that's not a promise.

The key aspects are mutual respect and that the boss still gets to be the boss. This is an I'm OK, You're OK interaction from the classic book of that name. Of course, mutual respect might be thin on the ground. In this case, techniques such as 'broken record' and 'paraphrasing the message' can be used to show you've listened and understood, without agreeing with it. For example, imagine a really difficult client who is proving impossible to work with, even though there's little profit in the contract, and who will prove even worse if you tell him you're going to walk away from the next job he wants you to do. (They'll withhold payment for the first job for a start.)

Among other things, they warn you that putting the price up may cost you that contract, so you say, 'I hear you, price is key, and putting the price up may cost us the next batch of work. Thank you, I've noted that.' (They'll be thinking 'good', but you can keep the 'though it's utterly irrelevant' to yourself until the time is right.)

Middle Bubble Training

Here's an example of a company that took that approach to non-technical skills and made it the centre of the strategy. It was called 'Middle Bubble Training' after transactional analysis (TA) theory, which says we can be in one of three states when interacting with others:

1. parent;
2. adult; or
3. child.

The child state is only appropriate in a brainstorming session, but the parent state should also be avoided as much as possible. The style of an authoritarian parent shouldn't be confused with a directive leadership style, which is appropriate when employees are inexperienced or at risk but often constitutes overuse of a top-down 'because I say so' mode. At its worst, an authoritarian style will engender a 'balancing' negative response, which we've discussed above and which is a disaster when trying to deliver an empowered and dynamic culture. However, even the more benign 'nurturing parent' style has two negative side effects.

First, if you assume you know best and are talking down to someone, it will inevitably hinder your listening. Second, any sort of paternal or maternal mindset won't do the worker's empowerment any good.

The 'adult' state in this case is simply about assertion, active listening and giving good feedback, especially when that is negative. Helpfully, the model lends itself to a user-friendly triptych, as illustrated in the 'snowman' diagram below. It's very easy to think of things in threes: so, top bubble, middle bubble or bottom bubble? And the key principle is simple: the more often employees are in their middle bubble, the better.

Unless you are a trained killer or some sort of sociopath, it's really difficult to engage that prefrontal cortex, think, be in the middle bubble and react simultaneously. For most of us, they are mutually exclusive, and many of us blurt out something without thinking that we instantly regret!

Here's a true example of how user-friendly the model is. Twenty years ago, one of the authors still occasionally frequented places where people congregate to drink alcohol and listen to music. Late one night, a huge man tapped the author on the shoulder and said, 'You just spilled my pint.'

'I don't think so,' the author replied, sizing him up. 'Yes you did,' he said, turning square-on, leaning over intimidatingly. Trying to stay calm, the author said, 'Look, I really haven't bumped into you or anyone else. You've got the wrong bloke'. Looking the author in the eye, too close for comfort, he insisted, 'No, I've got the right person,' but then with a giggle, 'but luckily for you I'm in my middle bubble and all's good! Ha . . . had you going there!' Thankfully, the author recognised a trainee from a behaviour-based safety course he'd run a few years before; a 6-foot 4-inch forklift driver with profound literacy issues. We mention this man's literacy issues only to stress how easy and appropriate it is to train everyone in interpersonal techniques that can sound complex in a textbook.

Such generic skills facilitate the effective flow of analysis and communication, both of which are utterly central to creating an empowering culture. They clearly fit under the 'nudge' banner, as they cost little but can have a big impact on behaviour.

If we really want to create a culture where 'it's OK to be challenged and it's OK to challenge', we need to equip individuals with the skills to do that.

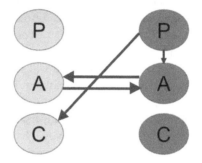

Keeping in your middle bubble by staying calm will very often lead to the other person calming down and moving into their middle bubble.

Matching and leading

The idea is that you meet your colleague (especially if they are some sort of 'opponent') halfway, then lead them to where you want to go.

So, for example, a naive but enthusiastic person bursts into the office ranting about a 'brilliant' idea (that's unworkable). What you do is leap up (rather than just sit there calmly and/or looking annoyed) and enthusiastically welcome them (with vigorous handshaking and other energetic behaviour similar to theirs), before leading them to a table where you can sit and from where you can slow the pace and volume of the conversation and take control of the discussion.

Now you can gently have them uncover the lack of practicality with your questioning coaching techniques ('It would be great if we could get to X, but how would we overcome Y?') and, in closing, politely request that as well as thinking things through more deeply they might knock next time they come to the office.

The matching senses approach from NLP (I see, feel, hear, etc. what you're saying, depending on which sense you first referred to) always makes us feel a bit uncomfortable, but the underpinning idea of avoiding a clash of style is one that comes quite naturally to empathetic people, with a clash in tone nearly always leading to a clash in objectivity and efficacy.

Coaching revisited

If you understand the feedback fish, you understand the basics of coaching. Imagine your four-year-old has brought you a picture of a fish but it's nothing better than an outline. Instinctively, you wouldn't exclaim, 'That's rubbish, it looks like a biscuit, you're wasting my time here.' Instead, you'd say, 'Oh that's fantastic . . . what a great fish. Thanks.' Then, because you want them to improve, you'll ask, 'I wonder . . . how could we make this picture even better? Let me think . . . how do fish see?' and the four-year-old will shout, 'An eye! We need an eye!' Then they'll add one in. In response, asking 'Fish swim, don't they?' we'll get fins and finally a fully formed fish.

It's the same principle with any coaching conversation. Even though they know you know, so long as they say it first they will have proved they knew it and will own it. Studies with electrodes monitoring brain function show it lighting up like a Christmas tree. We just don't light up when we say yes in agreement to a suggestion or recommendation from someone else. We may agree but we don't 'own it'.

A trap we can all fall into only too easily, and especially if we are a natural 'monitor evaluator', is to augment their idea with one of our own. Even though often it'll improve the idea by 5% or more, it'll almost certainly reduce their ownership and motivation by a full 50%. So if they don't ask for input, don't give it. Stick with, 'That's great, how did you come up with that?' and if you have an even better idea, keep to questions. Unasked-for advice is so often what's called 'positive but destructive'. It's better than the actively negative 'that'll never work' or passively negative 'it could work, I guess, if this doesn't happen', but none are as good as getting your report to say it first.

You know you've got it wrong when you find yourself commenting later, 'I was only trying to help!'

Motivational interviewing

This is coaching in its purest sense, as no advice is given at all, not even any specific leading questions. It all comes from the person, and it's all the more impactful for that. The rules are to ask:

- What would you like to improve on?
- Why would you like to do that?
- What's stopping you from doing it?

As above, the golden rule is to show patience and not give advice at all so as to engender a discovered learning/internal promise moment (as above, the best kind, as by far the most impactful). It will sound something like, 'Yes, I see clearly now what I must do, I can't believe I've let X put me off – out of my way, I'm a woman on a mission'.

'1 in 10' technique: a variation on motivational interviewing

A builder of submarines won a safety process of the year award by having a team of six tour the site and ask this question of a worker once a week. It's taken from educational psychology, where the counsellor tries to build on the answer from a terrible student, 'I'd score myself 1 on a 1–10 scale', by following up with the question, 'How can we get that up to a 2?'

In coaching, when someone has said, for example, that they are an 8/10 at driving safely, the key is to ask why they aren't a 0, rather than why they aren't a 10. Then when being told about all the good things they do we can nod and smile, provide positive feedback and generally build rapport. We should have, by this point, minimised defensiveness and be primed to have a constructive discussion. So we can also say, 'My job as a safety coach, seeking a step change, is to get you from 8/10 to 9/10 and halve those unsafe behaviours.'

In general terms, the same process can be applied to any element of wellbeing or to task re-engineering itself. This may sound trivial, but effectively applied it can transform your business team by team and task by task. For example, an old friend and colleague, Jim M, oversaw the doubling of turnover of his construction SME by applying the following system – looking first at stacking of materials in the engineering workshop and the loading and offloading vans. His system was very simple:

- First asking what goes well
- Then asking for a score of 1–10
- Then asking how they could push that score up towards 10.

Each one made the job a little quicker and easier to do, but the real benefit, he suggested, was a massive increase in motivation and engagement. He bought us a pint and proudly showed us pictures of pieces of kit that looked like giant toast racks and giant trays, all made by his engineers in their workshop. He conceded, 'They may look a bit rough and ready, but they're perfect for our guys' needs.'

It really is as simple as that.

Engendering ownership

You coach and listen but there needs to be clarity of thought between empowerment and egalitarianism. 'Here's a pot of money to spend, so do as you wish' will certainly engender ownership of the solution, add to the possibility of discretionary effort generally and is certainly appropriate in certain circumstances, such as the choice of safety footwear. However, it should never tip into an abdication of responsibility. A perception that this is so will, for example, reduce the possibility of discretionary effort generally.

An excellent example of genuine ownership of a behavioural-based safety process would be a shipbuilding company that offered a weekend for two in Amsterdam for a winning logo design. Many entries were received, and the prize was won, but the workers weren't impressed. 'Every day this ship is delayed costs them millions. What's a weekend for two in Amsterdam?' complained one employee, while others went further, alleging that the incentive was 'just a bribe, really'.

However, it was noticed that 11 entries were from children, and with a twelfth quickly 'commissioned' from the CEO's son, we had a calendar. This needed reprinting twice. With 2,000 workers on top of each other trying to get the platform finished, the job and the deadline were challenging, and the ship was only finally commissioned while being towed to its initial mooring, but there was not one single lost time injury (LTI).

Ownership comes primarily from choice and involvement. If you say it first, you own it, even if everyone knows you were led to the statement. So it's about the psychology of the leader-member exchange, and it's about perception. Management can't tell workers they have ownership; workers tell management. This is as much to do with one-to-one exchanges and soft skills as it is about delegation, roles and responsibilities.

Creating a reassuring environment

We'd hope that everything in this book would help to an extent, but it's worth covering a few key techniques from the world of influencing skills. In this context, they're not about influencing people to do what we want them to do but to set a tone of competence and integrity. They are:

- First, the one you heard when you were about knee-high (to quote the song). The basic polish your shoes, stand up straight, look the world right in the eye and speak clearly.
- Speak decisively. Don't say 'if I can' or 'I'll try' unless you absolutely have to.
- Never, ever say, 'I've been told to come and tell you . . .' (Either do it right or say no to the person who tasked you with passing the message on!)
- Treat people like adults by using data and illustration to make your case. We've all resented being told 'because I say so' since we were eight. (That people who are treated like adults tend to act like adults, and vice versa, 'theory X & Y', is perhaps the most famous of all self-fulfilling prophecies.)
- During dialogue, keep a neutral and open body position. Palms open, hand gestures natural and gentle. No folded arms, no pointing or chopping.
- Remember that humbly saying 'you're the expert' sets you up to be trusted, as you are at your most persuasive just after you've admitted a weakness (and trust development is our number one aim).
- Use the 'I' word, and get the person you're talking to to use it. We are something like three times less likely to break a promise if we use the 'I' word in a sentence. (For example, if when visiting a beach you ask someone to please keep an eye on your bags while you take a swim and they look you in the eye and say 'I will', then you will go swimming. On the other hand, if you ask someone mumbles 'sure' while looking at the ground . . . you won't go swimming!)
- Use people's names. It's the sweetest sound in the world to even the humblest person.
- Almost everything, including perceptions of integrity, goes better with a little bit of touching. (No giggling at the back – we mean appropriately firm handshakes, appropriate pats on the back or shoulder taps.)

People really skilled at these that have no integrity, such as some politicians and other sociopaths, can still be really influencing, but they very often leave you with a sense of discomfort. You say yes, but you feel suspicious. It's better to use them to help convey that you actually mean it!

Wellbeing, motivation and the undercover boss

If you've ever watched a version of the international TV show *Undercover Boss*, you'll have noticed that the boss always learns a lot and nearly always develops more respect and affection for their workers. It's a nice user-friendly validation of the techniques referred to earlier. However, it's also a variation on the old truism that it's impossible not to have at least some affection for a person whose story you know, and that links back to the holistic nature of wellbeing and motivation that commenced this section.

Most programmes finish with a worker being promoted, and they always beam with smiles and announce they can't wait to tell their family. Often in these shows, the boss will go on to say, 'I'm just so thrilled to have someone like you working for me', and very often the employee will well up. More money is always really welcome after all, but that's a practical issue.

Just like emotive music, being valued as a person gets you straight in the primeval reptilian brain. In practical terms, it means you're not likely to be pushed out of the cave anytime soon and are probably safe. Considering that we upgraded our brains millions of years ago to include the prefrontal cortex, it's amazing how many thoughts, behaviours and emotions are still controlled by that original old laptop! Am I safe? Am I useful? Can I trust you?

A good wellbeing process will skilfully engage both instinct and logic.

'If I had unlimited resources – I'd not give everyone therapy I'd give all parents and bosses emotional literacy training . . .'

(Dr Susie Orbach – one of the world's leading psychotherapists)

Organisational strategy and tactics   **8**

In short we felt that the following strategy is applicable to any organisation:

*   Assess where you are now on the key themes highlighted in the book.
*   Assume the organisation can take a more objective approach to learning that ensures any blame is always fair.
*   Assume that the soft skills of management and supervision are suboptimal and seek to enhance them, not just with training but with coaching and mentoring too.
*   Assume that there are problems with uncertainty, resources, trust, knowing how tasks fit into the bigger picture (traction) and other issues that impact on engagement and discretionary effort. Proactively seek them out, analyse them objectively and address them.
*   (To fast track that) set up workforce teams to systematically look at areas that are causing frustration or disengagement. (As best as possible let them choose these areas for themselves.)

and of course . . .

*   Train up some mental health first aiders. (Because even if your organisation is perfect – and if you think it is it *definitely* isn't – it's a war zone out there.)

Since this book was first published PAS 3002 guidance has been released, so it seems appropriate to start this chapter with an overview. It covers five strands and suggests a 'plan, do, review' approach.

1. Capitalise on diversity and inclusion as an organisational strength.
2. Proactively support the physical and psychological health and wellbeing of workers.
3. Foster a work culture that offers strong, ethical relationships, a collaborative and communicative management style and an organisational culture in which learning and development are encouraged.
4. Ensure jobs are designed so that they offer meaningful work.

And . . .

5. Support good people management policies and practices, including procurement design and risk management.

It's a thorough, sensible and correct document, but like documents of its type it's largely descriptive. The authors very much hope that the earlier sections of the book above give colour, illustration and rationale to this guidance.

Several interviewees explained that many of the larger companies totally 'get' the win-win argument, while others and 'the vast majority of SMEs simply don't'. In particular, it is suggested, they struggle to understand the distinction between occupational health, occupational hygiene and wellbeing. That said, even those companies that 'get' the win-win may not have this vital resource element covered in the way that other key issues are.

Despite the list of desirable and profitable behaviours that engagement brings, mental health and wellbeing may well not be on the risk register and be subject to very different governance. It's true that some of these divergences are created by the need to maintain medical confidentiality and protect data and that programmes such as rehabilitation and return to work are not the general undertakings of the safety professional. But this does not mean that safety and health should not be part of the same management system.

With typical governance arrangements in place, safety functions will report on the more physical nature of the business in terms of incident rates, reportable events, enforcement interest and potential claims information. Forming part of the larger risk register, safety will undoubtedly appear at some point on the scale; but where will the mental health and wellbeing of the workforce be shown and just how will that be calculated? For example, a safety risk statement might say something like:

> An incident occurs over which the business has little or no mitigation leading to major enforcement interest, public attention and financial penalties.

In the explanation that goes with this, there would be an estimate of how close to that event the business is sailing and perhaps something of the way it might occur. What you have to then show in your processes and management system is just how unlikely that is because of the controls that you have in place. As Chris Jerman, formerly of John Lewis, says:

> It's distinctly possible that many organisations come at mental health and wellbeing from the wrong direction as it were. Very often the first course of action will be to put in place a reporting and rehabilitation programme for those affected. This will be supported by an educational piece for managers and supervisors on how to refer people into the scheme and on recognition of signs and symptoms.

He adds:

> Cash flow is vital to any business but for long term viability its people are equally important. Which begs the question why boards don't have CCO directors.

Often, therefore, there are lots of 'apples in reception and a bike-to-work scheme, and they think they've got it covered'.

The NICE guidelines are very thorough and include the eight-step suggestion. We've included them verbatim here. You can see them reflected in the approach taken by organisations such as Crossrail.

1. Raise awareness through routine communication channels, such as email or newsletters, regular meetings, internal staff briefings and other communications with all relevant partner organisations. Identify things staff can include in their own practice straight away.
2. Identify a lead with an interest in the topic to champion the guidelines and motivate others to support their use and make service changes and to find out any significant issues locally.
3. Carry out a baseline assessment against the recommendations to find out whether there are gaps in current service provision.
4. Think about what data you need to measure improvement, and plan how you will collect it. You may want to work with other health and social care organisations and specialist groups to compare current practice with the recommendations. This may also help identify local issues that will slow or prevent implementation.

5. Develop an action plan with the steps needed to put the guidelines into practice, and make sure it is ready as soon as possible. Big, complex changes may take longer to implement, but some may be quick and easy to do. An action plan will help in both cases.

6. For very big changes, include milestones and a business case, which will set out additional costs, savings and possible areas for disinvestment. A small project group could develop the action plan. The group might include the guidelines champion, a senior organisational sponsor, staff involved in the associated services, and finance and information professionals.

7. Implement the action plan with oversight from the lead and the project group. Big projects may also need project management support.

8. Review and monitor how the guidelines are being implemented through the project group. Share progress with those involved in making improvements as well as relevant boards and local partners.

That's all good advice, of course, but we would like to suggest the following shorter and more prescriptive model, as we find people do like a short prescriptive model they can disagree with to help clarify their specific needs.

1. Get senior management buy-in. It's probably more important than everything else combined.

2. Screen. Where are we in terms of physical and mental health? What's the general culture like? Have we got all the leading edge metrics mentioned in this book in place?

3. Address mental health. It's a problem for your organisation certainly, so train up a full quota of mental health first aiders. If you're big enough to be able to insist, then make your contractors also adopt the approach.

4. Enhance non-technical skills for all. Unless you're a hugely unusual organisation, this is an area of great opportunity for you.

5. Address specifics. Get a lead team of champions together and train them up with what they need (applied nudge theory, other human factors skills such as ABC analysis, project management if need be). Then have individuals volunteer to take ownership of different core elements (as described throughout the book). Get them the resources they need and any specific extra training, then set them off.

   One team might look to resource and roll out 'financial savvy' workshops, with another looking at nudging physical exercise and another at sleep and fatigue issues. A project might look at how to get more people (men especially) to contact employee assistance and how effective it is generally.

Projects might also look at the different elements of the job that combine to make that job 'good work', and an organisation will certainly want a team to help embed a just culture decision tree into the discipline process if there isn't one.

6.   Monitor, coach, support, embed. It's an ongoing, dynamic process, and your project champions will, of course, be busy with it forever.

And so on. It's endless – just like the continuous improvement process.

Many organisations cannot, for legitimate operational reasons, put health and wellbeing at the top of their agenda and also continue to thrive themselves. There are three factors that dominate nearly all research into causal issues:

- regular work hours with no possibility of flexible hours or homeworking;
- long commutes; and
- shift work.

However, despite the fact that many organisations can realistically do little about these, that doesn't mean they should be fatalistic. Indeed, the opposite is true. They must proactively address the issues raised in this book and make best efforts to minimise their impact.

A 'culture of care' is vital. Admitting that a primary driver is sustainability is fine from within a win-win position. When we seek to improve productivity, absenteeism, 'presenteeism', 'leavism' and turnover for sustainability if it isn't genuinely also about the people *now* then scepticism and outright cynicism will most likely abound. And that very likely means no discretionary effort and limited gains for the business.

Allowing staff to undertake the various aspects of the programme in work hours is a key indicator; making access as convenient and pain-free as possible is essential as management commit to the process. The authors attended a presentation by the head of wellbeing of a large and well-known organisation. The woman went through a very impressive holistic model and then explained that the key focus was on 'wellbeing days'. These impressive-sounding events, she explained, were held on the weekend, not on work time. Not just for cost reasons, you understand, but because a key element of any self-improvement is committing to it, so this was, in effect, when you think about it, simply an opportunity to facilitate a 'sunk cost' motivation in their staff.

Someone in the audience announced a rude comment loud enough for her to hear. She ignored it, and we strongly suspect she's had practice at that.

Health screening

Mount Anvil screened all their employees in 2012 and found that 90% had high blood pressure. This led directly to an effective healthy eating campaign that meant that when readings were rechecked several months down the line there was a significant improvement in nearly all of the 90%.

Significantly, readings of one of the employees were so alarming and dangerously high that they had to tell him, 'Sit still, don't move!' and called an ambulance. They're pretty sure this saved his life on the day.

Almost every organisation that has run a thorough screening programme has a 'just in time' anecdote. In another case, a man's unusually high readings prompted extra tests that led to an early diagnosis of cancer. The treatment required was urgent but not especially onerous, and the individual took only a few days off work. A later diagnosis might have been fatal. The world of professional sport, with its screening programmes, is full of such stories.

Karl Simmons of Thames Water has suggested that the cost per person of a decent health screening is around £50 per person. He compares this with the cost of putting a car through an MOT, with the observation that staff are rather more valuable than ageing bangers!

Vitality run a free programme that more than 400 organisations have been through, where the organisation gets a report and free consultation, and individuals, following a 20-minute online assessment, get a 'vitality age', which, of course, proves fascinating to most people.

Organisational screening

MIND suggest that as well as a review of a variety of policies (see below), organisations should assess themselves across several dimensions as well as the self-evident ones: people management, preventative initiatives, employee support and job design. These, briefly, are:

- Senior leader buy-in and accountability. Is wellbeing clearly identified in a risk register with all the governance follow-up and assessment that should flow from that?
- Physical work environment – daylight, fresh air and an absence of stressors such as noise. (Incidentally, it's interesting to note the many organisations we have seen that make sure their computers have perfect heat and air

quality. Increasing but still small numbers are now also focusing on the same for their human resource.)

- Building mental health literacy. (See MH first aid advocacy as discussed above).
- Lived experience leadership. Increasingly, organisations are understanding the power of story and testimony generally, and there are consultancies dedicated to this issue.

It's been stated above that perhaps the key issue when we talk of mental health specifically is to destigmatise it and to make it 'OK to not always be OK'. It's very difficult to overstate the importance of lived experience testimony from senior management when it comes to achieving this.

Training and development

Some organisations we spoke to described up to 12 days of training/one-to-one coaching in batches of two days, and felt that each day or one-to-one session was useful and that, no, it didn't feel padded. Less senior managers got less training than this but still an extensive package. Others described really intensive sessions where their personality, values and integrity were put under intense scrutiny.

There are even three-week monitoring programmes, where (typically *senior*) managers are tutored in nutritional advice and fatigue management and are then monitored extensively so that the one-to-one feedback includes the fact that, 'You may not have felt more alert the day after that good night's sober sleep on Tuesday . . . but you were!'

All respondents suggested that following these, some people didn't buy in but that enough did for the initial inertia to be overcome, and those that assumed nothing would change have been influenced by the fact that things *did* change. This is the classic culture change model we so often see in the world of safety, where a tipping point is reached and a cynical individual becomes something of an outlier who stands out from the norm in a bad way.

This is a key area where senior management commitment must be strong and the follow-up well-coordinated. An excellent course is only base one; around 80% of the effectiveness in the medium to long term is delivered by the way the learning objectives are followed up, supported and embedded. If different departments are not united or if senior management commitment is at all tepid, then any organisation is primed to default back to the status quo.

Changes to work practices

Often a flexi-hours programme can help employees deal with stresses and pinch points outside of work. Trust is vital here, stresses Clive Johnson in interview with the authors. 'When there's mutual trust, then we don't have to worry about someone walking out of a door at 4.30 and we don't have to have a culture of not leaving before the boss does.'

The Tideway organisation, for example, has a metric that measures the amount of flexitime that employees are taking. There might be a very good operational reason why scores are low, but managers will be challenged on it.

We've discussed above that good work is good for you but bad work bad for you. Linking back to the screening issue discussed just above, organisations must remind themselves that it's not just a case of are their people fit for work but 'Is the work fit for their people?'

Organisational policies

MIND suggest a comprehensive and holistic list of proactive polices that need to be in place to underpin a strong wellbeing approach. These cover Health and Wellbeing, Flexible Working and Health and Safety, as you can imagine, but also:

- Bullying
- Discipline (see 'Just Culture' above)
- Sickness absence and return to work
- Performance management
- Change management
- Equality and Diversity

Although this isn't the place to go into the ins and outs of each, it's worth agreeing that this holistic and through approach makes sense. We know that uncertainty and change are always near the top of 'most stressful' lists and that holistic and progressive approaches tend to yield the best results. One simple example: some organisations make a point of counselling not just those who have been bullied but also those who did the bullying – recognising as a first principle that very often these are damaged people passing on their own hurt.

Valuing diversity is the first of the five core principles underpinning PAS 3002. A good practical example is that of the mental health champion Martin Coyd of MACE. One of his priorities is that of LGBT workers and based the data that in his industry, construction, LGBT workers are far more likely to suffer mental health issues. His success should not be surprising. Just like Gareth Thomas's acceptance by the rugby community and the NUM (Mineworkers) leading the 1985 Gay Pride march, LGBT concerns about expected prejudice were based on mostly unfounded assumptions. Maybe it's just easy to not feel the need to posture in a macho fashion when you *are* macho!

Changes to the bid process

When companies award contracts to other companies they are obviously in a very powerful position of influence. Again, we've learnt from the world of safety that all may not be what it seems. The conversation might go something like this:

> *As safety is one of our key values, we insist that you have this pennant/that award to even get yourself on the tender list.*
>
> *That's no problem – we do.*
>
> *Excellent, let's talk business!*

Anyone with any experience in safety knows that having a certain certificate on the wall might not mean very much in reality. Indeed, the working rule of thumb is that it's just wallpaper, and we only notice it when it's not there!

So, though better than not being there at all, this isn't much use when we're striving for excellence. However, as well as allowing a fuller understanding of the contractor's efforts, a proper in-depth discussion of a wellbeing programme sets a tone of expectation.

In the soft skills section we reiterate the fact that the words spoken typically make up a very small element of the overall message. The same is true of all interactions. When you mean it, people know. It's not just how many questions you ask; it's what the questions are, how you ask them, how you listen to the responses and how you react to the answers.

More specifically, companies are now insisting that contractors have dedicated 'first aiders' on-site, not just trained by the St John's Ambulance service but trained by MIND.

Though reflecting a snowballing trend, St John's Ambulance has recently joined the list of those who offer mental health first aid training too. It would be wrong to say that 'everyone is doing wellbeing and mental health now', but lots of organisations are. Quite possibly, several of your competitors. (Despite more and more organisations entering the market in mid-2018, Morgan-Sindall had so many volunteers for such training that they report they were struggling to find enough providers.) If you're not and that doesn't worry you and/or make you feel a little guilty then something has gone badly wrong with this book!

A final example of best practice

Earlier in the book we reported a finding by Vitality that internal reporting to the board about wellbeing metrics was around five times as important as having a wellbeing budget. We know that people are the solution, not the problem; we also know that most organisations that are world class achieved this by introducing even more and better rules than the ones that got them to a place where diminishing returns had set in. Instead, having realised they were at that fork, taking a proactive, learning-focused and person-centred approach was the only way a step change could be achieved.

A couple of years ago, one of the authors ran a senior management session with an SME that seemed to go quite well. The CEO said he recognised that they were exactly at that fork in the road and that the ideas they discussed were 'game-changing'. He marched straight out of the room, forgetting to say goodbye, but the head of SHE reassured the crestfallen-looking author that this was nothing personal and a good sign, as 'he does that when he's lost in thought'.

Returning recently, the author was struck by the changes. The whole place felt different: lighter, more energetic. The receptionist couldn't have been friendlier, and sitting in reception waiting for his contact to collect him, the author noticed a sign on his desk. It said 'Director of First Impressions'.

It was, of course, indicative of change. Those thoughts the CEO was lost in were good ones!

Developing a wellbeing programme: a step-by-step guide

Leadership

The first step when developing a holistic approach to wellbeing is to get top management properly engaged. They set the tone for the whole organisation, so they need to understand the importance of wellbeing and be committed to driving improvements.

Traditionally, health and safety has been driven by legislation. This has achieved a significant improvement in performance. However, rules will only ever lead to compliance. If we want to go beyond compliance and start to address issues such as wellbeing, then a different approach is required. You have to inspire people to get involved in continuous improvement and a drive towards excellence.

So how do you inspire people? Well, that's where leadership comes in. If you think about important historical leaders, there is no doubt that they were inspirational, but it's not just about what they said; it's what they did that made a real difference. Human beings are tuned to model behaviours; it's how we learn and develop from the earliest age. So to address wellbeing effectively, we need more inspirational leaders.

It's not a hard sell. The business case for managing health and wellbeing is well established, and very few business leaders would argue with the view that everyone has a right to go home fit and well every day. But to make a real difference, they need to have the confidence and commitment to demonstrate this in their everyday actions and interactions across the business. This will then be reflected through the management hierarchy and will help to establish a positive attitude to wellbeing within the organisation.

Engagement

The next step is to get the workers on board. People aren't comfortable talking about health and wellbeing matters, so without positive engagement and a shared sense of purpose, wellbeing initiatives are often viewed with suspicion and resisted by the workforce. So reach out to trade unions, worker representatives and volunteers, and form a working group to help drive change. They will be a valuable source of information and will play a vital role in communication and establishing a positive attitude to wellbeing within the organisation.

'It's estimated that up to 80% of all employees off with a 'bad back' are actually off with stress/MH issues' . . .

(Dame Carol Black, Fleming 8th Annual HSE Excellence Conference Key Note, Vienna, May 2014)

A MIND study in 2014 found that 95% who have taken work off due to stress feel unable to give their employer the real reason. Those with similar back aches but who are engaged with and positive about their work are the ones who ask in meetings 'is it OK with everyone if I stand up, walk around and stretch out for a bit? . . .'

Typically, the Safety Department, who are tasked with reporting directly to the board – because of the immediate importance of the safety metric – will be following this up by looking at manual handling issues, but it's very likely that the Occupational Health Department, who seldom report directly to the board, won't be looking at this at all. Nor will there be much, if any, interface between the two departments about the subject.

Shared purpose

When working to deliver change, it's important that all parties have a common understanding of the issue, the objective, the process and the timeline. So why not start by developing a policy statement or a manifesto for wellbeing? This should set out the organisation's commitment, values and objectives to help set the tone for the programme. It's also important to agree roles and responsibilities, as this will establish a delivery structure too.

Evidence base

To be effective, a wellbeing programme needs to address the issues that are of greatest relevance within the organisation, so it's worth reviewing stored data to identify any trends.

Sickness absence data, often held by HR departments, might point you to common ailments, or patterns of illness. Occupational health teams will also hold data that might be useful, and if you have an employee assistance programme (EAP), the provider will be able to share information about common issues raised and followed up through them.

Remember, there are confidentiality issues associated with health data. You will not able to access any information that could enable you to identify individuals. But you don't need this level of detail to inform a wellbeing programme. Grouped anonymised data that highlight general trends are all you need.

The Britain's Healthiest Workplace programme is a free service that helps organisations to measure health and wellbeing and to assess the effectiveness and value of wellbeing initiatives. There is no cost involved. The organisation and individual staff members are invited to complete confidential online questionnaires. The data are then analysed by experts and reports are sent back. The organisation gets a generalised report based on all the data submitted from the organisation. This helps to identify potential issues and to inform decisions about wellbeing and health promotion activities. It also benchmarks them against similar organisations by size and sector. This can be helpful in quantifying the case for change and measuring improvement once a wellbeing programme is established.

Staff members get a confidential report based on their own data, which estimates their 'wellbeing age', compares this to their actual age and provides information to help them make healthy lifestyle choices.

Targeting

The next step is to identify the issues that are impacting on wellbeing within the organisation to inform development of an improvement programme. Without this step, organisations can end up spending time, money and effort on initiatives that have little positive impact while more significant issues remain unaddressed.

There is a wide range of interlinked factors that can affect wellbeing. The overview holistic diagram in Chapter One illustrates key areas of focus, and you need to consider the factors within your organisation that might impact in each of these areas. What you're looking to create is a bespoke wellbeing needs assessment for your organisation. This can be really helpful in setting out the case for investment in a wellbeing programme. It will also inform development of your action plan.

A standard risk assessment model is ideal for this task. A simple five-step-type approach works well: identifying issues, evaluating impact, identifying and appraising existing controls and informing development of an action plan.

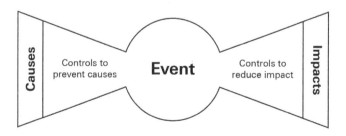

However, a bow tie approach is also very effective, creating a visual representation of the issues, impacts and controls (or barriers). This can be particularly effective if you have a working group in place, as it allows everyone to get involved in compiling the needs assessment, evaluating existing controls and identifying areas where new initiatives could add value.

Other similar methodologies can work well too, so select the one that will fit best within your organisation. Remember to keep the assessment at an organisational level, though. Wellbeing is a complex area, and the detail of the issues and impacts will vary enormously at an individual level. The intention is to identify broad aspects of work that could impact on wellbeing and to consider how these impacts can be managed at an organisational level. The detail of more specific issues and impacts at a personal level should be managed through established processes, with advice from specialists such as HR and occupational health professionals.

Prioritisation

As we mentioned earlier, there are a huge number of factors that can impact on wellbeing. Having mapped the factors that are impacting on wellbeing in your organisation and evaluated these, you need to prioritise areas for action. It won't be possible or effective to tackle everything all at once, so you need to target activities that will deliver the greatest possible impact and benefit.

A simple effort and impact grid can be really helpful here, particularly with a project team or working group.

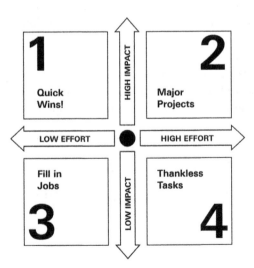

Interventions

Once you've identified the issues that are a priority in your organisation, you need to consider what action you can take to address these. There is a vast network of products and services available, but try to make sure that you remain focused on the needs of your organisation. Remember, expensive packaged solutions aren't necessarily always the best ones. You need solutions that are suitable for the people in your organisation and accessible to them. There's no point buying a fancy app if you don't allow the use of mobiles in the workplace. Salad options in the canteen make no difference to field-based staff!

Try to define the nature of the problem and then work with your project team to develop potential solutions that will work for your organisation. It's really important to get workers from all areas of the organisation involved at this stage of the process. People are much more likely to adopt something if they, or their colleagues, have had a role in developing it.

Resourcing

Once you've identified the actions that you would like to take, you will need to make a case for the necessary resourcing. The needs assessment, prioritisation and evaluation of potential interventions will help here, but you will need to follow the relevant company processes to secure the necessary funding, resources and organisational engagement.

Suppliers, contractors, charities and not-for-profit organisations may be able to support you. You will find that many of the issues you identify are common to many organisations. It's worth using your professional networks and searching online to find out what other people have done. You may well be able to learn from their experience. They might also have resources that they are willing to share.

It's also worth finding out if any of your colleagues have skills that they are willing to share to assist you. For example, sports or fitness qualifications, complimentary therapy or listening skills, nutritional knowledge or artistic abilities can all help to support your wellbeing programme, and often people welcome the opportunity to get involved.

Planning

You will need to develop a detailed project plan for your wellbeing programme. This will need to be realistic and time-bound, setting out the key steps for implementation of each of the initiatives and the actions and resources necessary to deliver these. Your working group or project team can help to develop this and use it to ensure that delivery stays on track.

There are lots of resources available online to support project planning, and you may also have departments within your organisation who can assist you with this.

Implementation

Once your plans are all in place and agreed, you can get on with implementing them. It's important at this stage to get as many people as possible involved in the programme. It's important to remember that the people who come forward as early adopters are likely to be those who are already comfortable with health and fitness matters. You need the programme to get to the 'hard-to-reach people', the ones who will hold back and appear sceptical at first, because these are the people who are likely to benefit most from the programme.

Your objective is to create a positive wellbeing culture in the organisation. If you haven't seen the YouTube clip about creating a movement, then it's well worth a look: www.youtube.com/watch?v=fW8amMCVAJQ. It stresses just how important the 'first followers' are to change processes. Just because it wasn't your idea doesn't mean your input isn't utterly vital. Don't forget the most successful captains in sports history are proud to be 'water bottle carriers'.

It won't necessarily happen quickly, so you'll need to be patient, but if you can get people involved, it will gain momentum as time goes on.

You need to use as many different approaches and techniques as you can. Trade union reps, staff reps and other bodies such as social committees and sports teams can help. Don't underestimate the benefits of competition either. Lots of organisations have successfully promoted an active lifestyle by giving all their employees pedometers and running an online league, which allows teams to compete against each other to achieve the greatest distance walked and win a donation to their chosen charity, or to be the first to cover the distance to the top of a mountain or a foreign country. This can work well but won't necessarily engage the hard-to-reach people. Peer pressure can be helpful, but it can isolate people too. You need to adopt a variety of approaches to get as many people as possible engaged in the programme.

Remember that wellbeing is not confined to the workplace. The idea is to promote thriving in every aspect of life, so reach out and involve families, friends and social networks too. It's much easier to engage in lifestyle change if you are supported by the people around you, and for some, being challenged, supported and encouraged by their loved ones can make a real difference.

Community groups can be a huge support too, and there are real benefits associated with bridging the gap between life and work.

You can work with your project team or working group to select a range of approaches that will fit your initiatives and the culture of your workplace and community. Keep reviewing progress, and if something doesn't seem to be working, stop. Take an objective look at it and think about what you could do differently to make it successful.

Metrics and measurement

You need to make sure that you will be able to determine whether each of your initiatives has been effective. To do this, you need to establish a baseline and then identify metrics that will allow you to monitor change. This might relate back to the data you used to identify the issues in the first place; you could monitor involvement or take-up of particular services or use surveys, etc.

It is important to define the following issues right at the beginning of your planning process:

- problem statement;
- stakeholders;
- current position;
- what the objective is;
- how long you expect it to take to achieve the objective, and any key milestones along the way;
- how you will measure progress;
- how you will report progress to stakeholders; and
- how long the initiative will last, and if/how it will transition to business as usual.

Short-term finite initiatives can be really effective in delivering change; however, you are aiming to deliver a sustainable wellbeing culture, so it's important to be clear about the long-term outcomes and how they can be maintained to deliver lasting impact and change ideas and behaviours within the organisation.

Comms and engagement

Wellbeing isn't something that you can 'do to' people. It can only be established through a shared sense of responsibility, accountability and engagement. It is therefore important to get the whole organisation committed and involved in the programme.

Culture change is explored earlier in this text.

One of the things that you might want to include in the programme is some basic training on conversation. Peer support can be very powerful, but, as discussed earlier, British people can be a bit reticent about discussing 'personal' issues.

Unfortunately, we tend to use phrases such as 'Alright?' and 'How are you?' as greetings rather than as invitations to engage. Most of us would be quite surprised and uncomfortable if people answered honestly! It can be hugely beneficial if you can reverse this trend and reintroduce some meaning to these phrases in your organisation.

The 'Time to Talk, Time to Change' programme has some great advice and free resources. See their website for more information (www.time-to-change. org.uk/about-us/about-our-campaign/time-to-talk).

Managing expectations

Culture change doesn't happen quickly. It's a long-term multifaceted process involving lots of different elements. As with any change project, it is inevitable that you will encounter problems with your wellbeing programme. Not everything will work out well, and it will be necessary to make adjustments along the way.

This is OK. It's a normal aspect of delivering change. But when you are trying something new with a number of stakeholders, small setbacks can have a significant negative impact. It's important to manage expectations at the outset. Help people to understand that you might not get everything right first try, but give them confidence in the monitoring system and help them to understand that this provides an opportunity to make adjustments along the way to ensure the ultimate success of the programme.

This can be particularly important with senior leaders. Try to get a senior sponsor for the programme who can be part of the working group; this will help with engagement, oversight and expectations management.

Remember to stay positive and to use positive language. There are no 'failures' just a lot of 'opportunities for improvement'!

Backup

Wellbeing naturally impacts on health, so it's important to make sure that you have a network of specialist professional backup that you can call on for support.

You may well identify people who have particular individual health issues that require treatment or specialist advice. It's important that such people are referred quickly to qualified professionals.

So, during your planning, identify a source of professional support for each area of wellbeing that you plan to address. Make sure that these services are briefed and that you are able to put people in touch with them quickly if required.

For many issues it may be relevant to direct people to their GP or local NHS services, but you should also consider how support might be obtained from your occupational health team and / or services such as employee assistance programmes.

Make sure that support services are mapped and that information is shared with everybody at every stage of the programme so that people are able to support and assist each other.

Conclusion

The eagle-eyed among you may have spotted a familiar approach here.

The process that we have described could be equated to the POPIMAR model or the Plan-Do-Check-Act cycle, which have underpinned the improvement that we have achieved in health and safety over the last 40 years. This is no accident. These processes have served us well, delivering significant positive change. They offer a structured approach and a feedback cycle that supports ongoing development and continuous improvement, and they are applicable to pretty much any change process.

We have achieved a very positive change in terms of safety at work. The challenge now is to replicate this for health and wellbeing.

It's not rocket science! People matter, and we have an opportunity to make a significant difference to individuals, organisations and society by applying the knowledge that we have gained in the safety field to help establish a sustainable wellbeing culture in our workplaces.

James Reason is perhaps the godfather of safety excellence, so perhaps it's apt that the last word goes to him. He suggests we treat safety like a guerrilla war. It's one that we can't ever win, but if we're smart and utilise the resources at our disposal cleverly we can fight a pretty good rear-guard action so that sometimes the invading army, like Russia in Afghanistan perhaps, gets exasperated and goes home. There's a powerful metaphor for life and wellbeing itself in there somewhere!

Good luck.

Glossary

Belbin, Dr Meredith – developer of the Belbin's Team Roles theory, which states there are nine different behaviours (or contributions) that individuals display in the workplace

Bradley Curve (the) – created by DuPont in 1995 to try and benchmark notions of culture and performance in relationship to safety

Breathe Freely – a BOHS initiative aimed at reducing occupational lung disease in the UK, which causes an estimated 13,000 deaths per year

Brexit – United Kingdom's withdrawal from the European Union as voted for in the UK referendum on 23 June 2016

Britain's Healthiest Workplace – an annual survey that aims to give recognition to Britain's healthiest employers and employees and to increase understanding around poor health and wellbeing among staff

Buncefield fire – incident caused by a series of explosions at the Hertfordshire Oil Storage Terminal, 11 December 2005

CALM – Campaign Against Living Miserably, a UK campaign dedicated to preventing male suicide, the biggest killer of men under 45

Campbell Institute – America's National Safety Council, center of excellence for environmental, health and safety management

Cascarino, Tony – former footballer who played as a forward for various British and French clubs and for the Republic of Ireland national team

CBT – Cognitive Behavioural Therapy

Dweck, Carol S. – Professor of Psychology at Stanford University

Erasmus Programme – European Region Action Scheme for the Mobility of University Students

Generation Y – the generation of people born during the 1980s and early 1990s

Gig economy – a way of working that is based on people having temporary jobs or doing separate pieces of work, each paid separately, rather than working for an employer

Glassdoor – a website where employees and former employees anonymously review companies and their management

Grift principle – refers to the Grit principle (see below); however, Grit is more about personality . . . grift is like graft and is used here to acknowledge the environmental context

Grit principle – GRIT is Martin Seligman's name for the well-known (to safety people) Heinrich Principle, which states that we tend to get out what we put in (though with no guarantees either way . . .) (see Chapter 6 – *Flourish*). This topic is also covered extensively in Matthew Syed's book *Bounce*

G7 – Group of Seven: the economic alliance of Canada, France, Germany, the UK, Italy, Japan and the US

G20 – Group of Twenty: a group of 19 individual countries plus the EU; these are the world's biggest economies, who meet regularly to discuss the situation in the world economy

Heinrich triangle – American industrial safety pioneer Herbert Heinrich showed that for every accident resulting in a fatality or major disabling injury there are approximately 300 unsafe incidents

HR – human resources

HSE – Health and Safety Executive, Britain's independent regulator for work-related health, safety and illness

HSL – the Health and Safety Executive's laboratory. It conducts scientific, medical and technical research in the field of health and safety

HSW Act – The Health and Safety at Work etc Act 1974 (also referred to as HSWA, the HSW Act, the 1974 Act or HASAWA) is the primary piece of legislation covering occupational health and safety in Britain

Hygge – a quality of cosiness and comfortable conviviality that engenders a feeling of contentment or wellbeing, taken from Danish culture

IOSH – Institution of Occupational Safety and Health, organisation for health and safety professionals in Britain

Koen, Dr Susan – organizational psychologist in human reliability and human fatigue in the workplace

Kronos Incorporated – US-based multinational workforce management software and services firm

Mates in Mind – framework for the UK construction industry to raise awareness and address the stigma of poor mental health

Mental Health First Aid (MHFA) – a training course which teaches people how to identify, understand and help someone who may be experiencing a mental health issue

Milgram, Stanley – a psychologist at Yale University who conducted an experiment focusing on the conflict between obedience to authority and personal conscience (1974)

Mount Anvil – London-based property developer

NASA – National Aeronautics and Space Administration

Neoliberalism – the 20th-century resurgence of 19th-century ideas associated with laissez-faire economic liberalism. These include economic liberalization policies such as privatisation, deregulation, unrestricted free trade and reductions in government spending in order to increase the role of the private sector in the economy and society

NICE – The National Institute for Health and Care Excellence, non-departmental public body of the Department of Health in the UK

Ogilvy & Mather – a global advertising and marketing agency based in London

ONS – Office of National Statistics

Outtakes – a film producing company specialising in health & safety films

P&L – Profit and Loss

PAPYRUS – Prevention of young suicide UK

Piper Alpha disaster – offshore oil and gas platform that suffered an explosion in July 1988, killing 167 workers. It is still regarded as the worst offshore oil disaster in the history of the UK

Player, Gary – ex South African professional golfer and winner of nine major championships

POPIMAR – acronym that describes the key elements of successful health and safety management. It stands for Policy, Organising, Planning and Implementation, Measuring performance, Audit and Review

Post-traumatic growth – a positive psychological change experienced as a result of adversity, a term coined by psychologists in America during the mid 90s

Reason, James – professor of psychology at University of Manchester, developed key theories on accident causation and is a widely published author

Sandwell Community Caring Trust – a charity that provides residential and respite care services across the Black Country in the UK

Schneider, Benjamin – proposed the attraction-selection-attrition (ASA) model for understanding organisations which says that personal characteristics of employees are likely to become more similar over time (1987)

Senna, Ayrton – Brazilian racing driver who died in a car crash while leading the 1994 San Marino Grand Prix

Severn Trent plc – a UK water company that is traded on the London Stock Exchange

SHE – Safety Health and Environmental body of knowledge

SMEs – small and medium enterprises

State of Mind – a charity that harnesses the power of sport to promote positive mental health among sportsmen and women, fans and wider communities

Stoddart review – makes the case for the vital role that the workplace can play in productivity, and shares best practice, leading opinion and data

SWOT analysis – a study undertaken by a person or organisation to identify strengths and weaknesses, external opportunities and threats

Thames Tideway Tunnel – a 25 km tunnel under the tidal section of the River Thames which when constructed will serve as 'super sewer' for the capital

Thames Water – largest water and wastewater services company in UK

Time to Talk, Time to Change – an annual day in the UK to encourage people to get talking and break the silence around mental health problems

Undercover Boss – British reality television series. Each episode depicts a person in a high management position at a major business who decides to go undercover as an entry-level employee to discover the faults in the company

Vitality – Health and Life insurance company based in London

Warr, Peter – workplace psychologist who developed a model called the Vitamin Model to describe the most common job characteristics that impact employee wellbeing and development

White Paper/s – thought leadership publications designed to stimulate discussion and debate

Zero hours – an employment contract which does not oblige the employer to provide regular work for the employee but requires the employee to be on call in the event that work becomes available

References and suggested further reading

Alter, A. (2013) *Drunk Tank Pink: And Other Unexpected Forces That Shape How We Think, Feel, and Behave*. London: Oneworld.

Bean, T. and Lang, A. (2017) *The Wealthy Body in Business*. London: Bloomsbury.

Black, C. (2014) Key Note presentation at the Fleming 8th Annual HSE Excellence Europe Forum 22nd May in Vienna, Austria.

Blanchard, K. (1982) *The One Minute Manager – Increase Productivity, Profits and Your Own Prosperity*. London: Collins.

Buckingham, M. (2005) *Now, Discover Your Strengths*. New York: Pocket Books.

Buckingham, M. (2007) *Go Put Your Strengths to Work*. New York: Free Press.

Buckingham, M. (2009) *Find Your Strongest Life*. Nashville, TN: Thomas Nelson.

Burrell, G. and Morgan, G. (1979) *Sociological Paradigms and Organizational Analysis*. Portsmouth, NH: Heinemann Educational.

Carnegie, D. (1936) *How to Win Friends and Influence People*. London: Vermilion.

Covey, S. (2004) *The 7 Habits of Highly Effective People*. New York: Simon & Schuster.

Ehrenreich, B. (2009) *Smile of Die*. London: Granta.

Frankl, V.E. (1959) *Man's Search for Meaning*. London: Rider Books.

Gladwell, M. (2009) *Outliers – the Story of Success*. London: Penguin.

Johnson, S. (1999) *Who Moved My Cheese? An Amazing Way to Deal with Change in Your Work and in Your Life*. London: Vermilion.

Kahneman, D. (2011) *Thinking, Fast and Slow*. London: Penguin.

Lanier, J. (2018) *10 Arguments for Deleting Your Social Media Accounts Right Now*. London: The Bodley Head.

Lencioni, P. (2002) *The Five Dysfunctions of a Team*. San Francisco, CA: Jossey-Bass.

Orbach, S. (2018) 'In Therapy' comment during a live 'interview' event with Jeannette Winterson, University of Manchester, 16th April.

Read, S. (2013) Quoting Shelter research on financial insecurity. *The Independent*, 10 April.

Robertson, I. and Cooper, C. (2011) *Well-Being: Productivity and Happiness at Work*. London: Palgrave Macmillan.

Seligman, M. (2011) *Flourish: A New Understanding of Happiness and Well-Being – and How to Achieve Them*. London and Boston: Nicholas Brealey Publishing.

Sunstein, C.R. and Thaler, R.H. (2008) *Nudge: Improving Decisions about Health, Wealth and Happiness*. New Haven and London: Yale University Press.

Syed, M. (2010) *Bounce: The Myth of Talent and the Power*. London: Fourth Estate.

Syed, M. (2015) *Black Box Thinking: The Surprising Truth about Success (and why some people never learn from their mistakes)*. London: John Murray.

UK Department of Transport (2018) 'Socio Economic Class and Cycle Use'. National Travel Survey.

Ulmi, N. (2016) 'Hygge, la curieuse histoire du bonheur danois', *Le Temps* magazine, 4 November 2016. Retrieved from: www.letemps.ch/societe/2016/11/04/hygge-curieuse-histoire-bonheur-danois.

Wilkinson, R.G. and Pickett, K. (2009) *The Spirit Level: Why More Equal Societies Almost Always Do Better*. New York: Bloomsbury Press.

YouGov ('Mind') Survey (2014) 'Non-Disclosure of Reasons for Absence from Work'.

About us

The British Safety Council is a not-for-profit membership organisation and a registered charity. We have over 5,000 corporate members in the UK and across the world. We were established in 1957 with the vision that no one should be injured or made ill at work, and for over 60 years we have campaigned for improvements in standards and attitudes to health as well as safety matters. We have an excellent reputation in the sector and a strong track record for identifying issues and driving change.

We first promoted health and wellbeing at work back in the 1960s, and this continues to be a core component of our agenda to the present day. For more information about the organisation, please see www.britsafe.org.

About the authors

Tim Marsh

Tim Marsh, formerly a post-doc at the University of Manchester, was one of the team leaders of the first European work in the practical application of Behavioural Safety techniques. He formed Ryder-Marsh Safety in 1994 and has worked as a consultant in the fields of safety leadership, safety culture and organisational culture generally with more 400 companies worldwide. Previous books include *Talking Safety* (2011) and *A Definitive Guide to Behavioural Safety* (2017). After closing Ryder-Marsh in 2015 to focus on family issues he founded the consultancy Anker and Marsh in 2018. He has been Honorary Professor at the University of Plymouth since 2015.

Louise Ward

Louise Ward is a Chartered Health and Safety Professional with over 17 years' experience in a variety of sectors, including nuclear power, newspaper production, investment banking, facilities management, manufacturing and the Civil Service, railway operations and waste water management.

Index